Food Science and Technology

nova
science publishers
www.novapublishers.com

Food Science and Technology

Nutraceuticals: Food Applications and Health Benefits
Anita Kumari, PhD (Editor), Gulab Singh, PhD (Editor)
2022. ISBN: 978-1-68507-488-3 (Hardcover)
2022. ISBN: 978-1-68507-512-5 (eBook)

Food Processing: Advances in Research and Applications
Myriam Huijs (Editor)
2022 ISBN: 978-1-68507-570-5 (Hardcover)
2022 ISBN: 978-1-68507-581-1 (eBook)

A Closer Look at Polyphenolics
Peter Bertollini (Editor)
2021 ISBN: 978-1-68507-434-0 (Hardcover)
2021 ISBN: 978-1-68507-480-7 (eBook)

Cheeses around the World: Types, Production, Properties and Cultural and Nutritional Relevance
Ana Cristina Ferrão, Msc (Editor),
Paula Maria dos Reis Correia (Editor),
Raquel de Pinho Ferreira Guiné, PhD (Editor)
2019 ISBN: 978-1-68507-434-0 (Hardcover)
2019 ISBN: 978-1-68507-480-7 (eBook)

Flour: Production, Varieties and Nutrition
María Dolores Torres Pérez (Editor)
2018 ISBN: 978-1-53613-761-3 (Hardcover)
2018 ISBN: 978-1-53613-762-0 (eBook)

More information about this series can be found at https://novapublishers.com/product-category/series/food-science-and-technology-series/.

Sanket J. Joshi,
Susmita Mukherjee
and Moupriya Nag
Editors

Proceedings of BIOSPECTRUM

The International Conference on Biotechnology and Biological Sciences: Biotechnological Intervention towards Enhancing Food Value

nova
science publishers
www.novapublishers.com

DOI: https://doi.org/10.52305/LAOH6077.

Library of Congress Cataloging-in-Publication Data

ISBN: 978-1-68507-985-7

Published by Nova Science Publishers, Inc. † New York

Dedicated to our loving family.

Contents

Preface

Since 2017, the Department of Biotechnology, University of Engineering & Management, Kolkata, India, every year organizes the International Conference on Biotechnology & Biological Sciences: BIOSPECTRUM. Experts from various institutes of national and international repute belonging to different parts of the world participate in this conference and share their interesting and novel results. Every year we had approximately 150-200 participants, and after careful review we select and finalise around 20-30 papers in each individual category for publications. BIOSPECTRUM: The International Conference on Biotechnology & Biological Sciences is a unique platform to network and collaborate with scientists and academicians working in the country and abroad. This conference provides an avenue for interactions and knowledge transfer among budding researchers, established scholars, and professionals from different divisions of biological sciences. Over the years, it has provided a platform to exchange ideas among the peers, thus supporting networking opportunities and collaboration, in India and beyond. During different sessions, it acted as a forum to initiate the dialogue between industry and academia and take it further. We, in this conference, try to bring forth the idea of 'Lab to Field', so that the researchers at the lab can be brought forward for human welfare. To strengthen this concept, we collaborated with NOVA Science Publishers, so that the selected and peer-reviewed papers can be published as conference book proceedings, which will eventually be widely publicised at national and international levels.

In the current book, Proceedings of BIOSPECTRUM: The International Conference on Biotechnology and Biological Sciences: Biotechnological Intervention towards Enhancing Food Value, we selected different manuscripts dealing with the theme of biotechnological intervention towards food value chain. These chapters cover the majority of topics of concern for scientific communities, researchers, and policy makers. Subjects discussed include biotechnological intervention in food security and value chain, practices to improve high quality agricultural produce and improved food security, probiotics, antioxidant activity of edible oils, fermentation to improve protein and phenolic profiles of plant products, marker assisted breeding for genetic improvement of pest-insect resistance in cash crops, and more. We believe this book will prove to be a vital source of information, touching all three spheres of biotechnological food research and applications.

The editors are grateful to all the conference participants, contributing researchers, and expert academicians, whose involvements, either as presenters, authors, or reviewers, made this book a possibility. We would like to express our sincere gratitude to our family members, who were always supportive during the course of such additional scholarly activities and apart from our routine academic duties, gave us the strength to complete the task on time. While we endeavored to make sure that this book is free from any errors or misleading information, any such mistakes are completely unintentional, and if you find any such mistakes, then kindly pardon us. We are thankful to NOVA Science Publishers team for their relentless support throughout the publishing process. We would also like to sincerely acknowledge our universities for extending the facilities and encouragement for such scholarly activities.

Sanket J. Joshi,
Susmita Mukherjee,
and Moupriya Nag

Chapter 1

Paternity and Maternity Determination of Pigeonpea (*Cajanus*) Interspecific Hybrid Seedlings Using RAPD Markers

Alok Ranjan Sahu[1,2,*] and Jogeswar Panigrahi[1,3]

[1]Plant Biotechnology Lab., School of Life Sciences,
Sambalpur University, Jyoti Vihar, Sambalpur, Odisha, India
[2]Department of Botany, Vikash Degree College,
Barahaguda Canal Chowk, Bargarh, Odisha, India
[3]Department of Biosciences and Bioinformatics,
Khallikote University, Berhampur, India

Abstract

Genetic relationships of the F1 hybrids and their parents *Cajanus cajan* cv. ICP26 and *Cajanus scarabaeoides* acc. ICPW94 as female and male parent respectively were analyzed with 20 RAPD Primer. Out of 20 primers, 17 (85%) primers showed polymorphism between the parents, which could be used for purity assessment of the hybrids. A total of 26 and 21 RAPD markers unique to *C. scarabaeoides* parent and *C. cajan* parent respectively, were also inherited to F1 hybrids. Further, only seven RAPD (OPO01, OPO02, OPO05, OPO06, OPO07, OPO12, and OPO13) primer pairs produced male and female parent-specific markers that were inherited to F1 hybrid also. These results demonstrate that these RAPD markers should be useful for the identification of F1 hybrids by using the same parental line.

Keywords: *Cajanus*, F1 hybrid, hybridity, pigeonpea, purity identification, RAPD

---

* Corresponding Author's Email: alok.btgene@gmail.com.

In: Proceedings of BIOSPECTRUM
Editors: Sanket J. Joshi, Susmita Mukherjee and Moupriya Nag
ISBN: 978-1-68507-985-7

Introduction

Cajanus cajan (L.) Millsp. is commonly known as pigeonpea and belongs to the Fabaceae family. Pigeonpea are one of the most important legume crops of semi-arid and tropical regions. It is a high protein legume whose seeds contain 20-24% protein. As per FAOSTAT, 2012; around 4.75 Mha of land were used for cultivation of pigeonpea, having a production of 3.68 MT. The significance of pigeonpea in our daily diet is well recognized. India is leading both in the area of cultivation and production of pigeonpea, but sincefrom the last five decades, the productivity has remained stagnant. There has been no major breakthrough in its production because pigeonpea is affected by both abiotic and biotic stress factors. A narrow genetic deviation withlong lifecycle of these legumes and heterozygous genome structure create its genetic development only for selection and maintenance of useful type of gene action [19]. Although the wild relative of cultivated *C. cajan*, has genesfor annuality or pernniality, earliness, photoperiod insensitivity, biotic and abiotic stress resistance [20]. *Cajanus scarabaeoides* accession of the secondgene pool has resistance gene to pod borer and its associated pests, flowering, photoperiod insensitivity, and high protein content [16-18, 20]. In addition, to improve pigeonpea cultivars these agro-economic charactersbearing genes might be inserted into the cultivated *Cajanus cajan* through hybridization technique, which will help to extend new genotypes.

The interspecific F1 hybrids could be the initial materials for making Advanced Breeding Lines and help us to study the relationship between both the genomes in a single cytoplasmic background. The successful production of interspecific hybrid between *Cajanus cajan* and *Cajanus scarabaeoides* was reported by a few researchers [2, 9, 16, 18, 26]. The identification of true hybrids at the plantlet stage is quite essential in inter-specific hybridization. In order to the identification of hybrids, phenotypic and cytological analysis was used in several crops including genus *Cajanus*. Due to environmental conditions and developmental factors, the capability of cyto-morphological markers has drawbacks in their uses; however, it is easy to identify the true hybrids in the seedling stage by using molecular markers. For the first time, Mishra et al. [9] reported on cyto-morphological and molecular characterization (RAPD and ISSR markers) of *Cajanus cajan* (ICP26) and *Cajanus scarabaeoides* (ICPW94) and their F1 hybrid.

Keeping in mind, several molecular markers were developed to add the usual less effective tools for molecular studies at the genetic level [12, 24, 25]. In general, RFLP (Restriction Fragment Length Polymorphism), RAPD

(Random Amplified Polymorphic DNA), ISSR (Inter Simple Sequence Repeat), AFLP (Amplified Fragment Length Polymorphism), SSR (Simple Sequence Repeat), SNP (Short Nucleotide Polymorphism), AFLP-RGA (Resistance Gene Analogue Anchored AFLP), SRAP (Sequence Related Amplification Polymorphism) markers were generally used for phylogenetic analysis and genetic diversity studies [1, 3, 4, 5, 11, 13, 15, 16] in pigeon-pea. Sahu et al. [18] constructed a genetic linkage map on interspecific F_2 lines derived from *Cajanus cajan* (ICP26) and *Cajanus scarabaeoides* (ICPW94) by using SCoT (Start Codon Targeted Polymorphism), RAPD, ISSR markers and markers and SITLs. In this present study, we evaluated the genetic purity of the F_1 hybrid involving ICP26 and ICPW94 using RAPD markers. Being the ICPW94 as the male parent and ICP26 as the female parent, the uses of these ICPW94 and ICP26 specific markers might be used to find out the paternity and maternity determination of Pigeonpea hybrid in seedlings stage and also the marker-assisted selection of germplasms during breeding programmes.

Materials and Methods

DNA Extraction and PCR Amplification

From ICRISAT, Patencharu, Hydrabad; *Cajanus* seeds were brought and maintained at the Experimental Garden of School of Life Sciences, Sambalpur University, Odisha, India. Screening of all the 34 genotypes for host resistance to pod borer and its associated pest were carried out [18]. As reported by Sahu et al. [18] the hybrids plants developed by using *Cajanus scarabaeoides* (ICPW94) were considered as resistant parent or male parentsand *Cajanus cajan* (ICP26) as susceptible or female parents. The modified CTAB method was used to extract the genomic DNA from both the parents and their hybrids [21]. The concentration and purity of genomic DNA were measured by using the UV-Visible spectrophotometer (UV 1, Thermo Electron Corporation, UK) by using $T_{10}E_1$ buffer as blank. For PCR amplification, the entire genomic DNA was diluted and equilibrated to 10 ng/μl in the buffer. For more conformation, the diluted DNA samples were analyzed on 0.8% (w/v) agarose gel electrophoresis along with uncut phage λ-DNA as standard.

For PCR amplification a total of 20 RAPD primers of OPO series (OPO01-OPO20) were selected and the amplification of PCR was carried out by using 25ng of genomic DNA. The PCR amplification conditions were the same as described by Mishra et al. [9]. The amplified products were run on 1.4% (w/v) agarose gel electrophoresis. The amplified DNA fragments were visualized and photographed on the gel documentation system (Fire Reader System, UVITECH, UK). In order to calibratie the size of amplified fragments, a 250 bp ladder (Bangalore Genei Pvt. Ltd.) was used as a standard. By using the TL-120 software (Non-linear Dynamics, Total Lab. Ltd., Newcastle Upon Tyne, UK) the molecular weight of amplified fragments were calibrated.

Results

RAPD Marker Analysis

Out of 20 RAPD primers, 19 responded including 17 polymorphic primers, which generated 148 unequivocal scorable fragments, out of the 85 (57.43%) were polymorphic at the parent level. Among the parental polymorphic fragments, 41 and 44 fragments were unique to the ICP26 and ICPW94 genotype, respectively (Table 1). The sizes of amplified fragments were ranged from 225 to 3110 bp. Forty-seven parental polymorphic fragments, including 21 female (ICP26) specifics and 26 male (ICPW94) specifics, were inherited by the F_1 hybrid. Thirty-eight numbers of amplified fragments were found in both the parents, but not found in F_1 hybrids. Seven primers (OPO01, OPO02, OPO05, OPO06, OPO07, OPO12 and OPO13) were able to show hybrid (25 markers) banding patterns between the parents and the F_1 hybrid (Table 1, Figure 1).

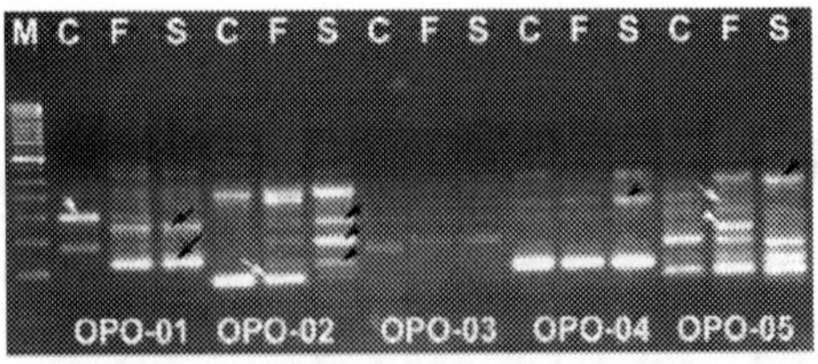

Figure 1. RAPD banding pattern generated by five OPO series primers showing inheritance of *Cajanus cajan* (ICP 26) specific (White arrow), and *Cajanus scarabaeoides* (ICPW 94) specific polymorphic fragments to F1 plants (Black arrow), Lane 'M' represents molecular weight marker; C: *Cajanus cajan*; S: *Cajanus scarabaeoides*; F: interspecific hybrid.

Table 1. Parental polymorphism analysis by using 20 RAPD primers, and inheritance of RAPD markers to the interspecific F1 hybrids raised involving *Cajanus cajan* cv. ICP26 and *C. scarabaeoides* acc. ICPW94

Primer	Primer Sequence (5'→3')	No. of Fragments Amplified	Amplified Fragment Size (Bp)	Praental Polymorphism (%)	No. of Polymorphic Fragments in Parents		No. of Parental Polymorphic Fragments Inherited to F1	
					Cc	Cs	Cc	Cs
OPO01	5'-GGCACGTAAG-3'	09	345-1805	55.5	1	4	1000	1635, 835, 590, 345
OPO02	5'-ACGTAGCGTC-3'	07	285-2405	57.14	1	3	2405	950, 760, 590
OPO03	5'-CTGTTGCTAC-3'	04	225-770	100	3	1	--	770
OPO04	5'-AAGTCCGCTC-3'	03	570-1615	--	--	--	--	
OPO05	5'-CCCAGTCACT-3'	11	525-2405	54.5	2	4	880	1520
OPO06	5'-CCACGGGAAG-3'	12	365-3110	58.3	3	4	1050	1460, 715
OPO07	5'-CAGCACTGAC-3'	11	250-2110	90.9	7	3	1620, 1265	1860, 990
OPO08	5'-CCTCCAGTGT-3'	07	420-1750	71.4	--	5	--	--
OPO09	5'-TCCCACGCAA-3'	06	580-1460	50	2	1	875, 760	--
OPO10	5'-TCAGAGCGCC-3'	05	795-2720	--	--	--	--	--
OPO11	5'-GACAGGAGGT-3'	12	475-2335	66.6	4	4	--	660, 475
OPO12	5'-CAGTGCTGTG-3'	08	475-1570	25	1	1	950	870
OPO13	5'-GTCAGAGTCC-3'	08	385-1830	75	2	4	1645, 550	1250, 1090, 385
OPO14	5'-AGCATGGCTC-3'	06	700-1860	66.6	3	1	1000, 850, 700	--
OPO15	5'-TGGCGTCCTT-3'	10	350-1570	60	5	1	1570, 1415, 1115, 880, 560	--
OPO16	5'-TCGGCGGTTC-3'	09	715-2945	55.5	4	1	1410, 1060	--
OPO17	5'-GGCTTATGCC-3'	Not amplified						
OPO18	5'-CTCGCTATCC-3'	07	500-1370	57.1	--	4	--	1370, 1220, 985, 810
OPO19	5'-GGTGCACGTT-3'	07	450-2445	57.1	3	1	--	1075
OPO20	5'-ACACACGCTG-3'	06	535-1665	33.3	0	2	--	1665, 635
		148	225-3110	57.43	41	44	21	26

Cc-*Cajanus cajan* (ICP 26) specific band; Cs-*Cajanus scarabaeoides* (ICPW 94) specific band.

Discussion

For all breeding programmers, F_1 interspecific hybrids are kept as starting points and also required to identify the F_1 interspecific hybrids at the seedling stage concerning extensive hybridization [7]. Similar kinds of results were found using RAPD and ISSR markers [9, 17]. However, in the F_1 hybrid, some of the parental amplified fragments were absent and a few non-parental amplified fragments appeared. This result occurred due to the recombination of DNA with minor genomic reorganization [9]. Further, due to crossing over during meiosis, the loss of parental amplified fragments in F_1 hybrids might be endorsed to defeat priming sites [9, 22]. Since the objectives here concern to finding out the paternity and maternity of Pigeonpea hybrid in the seedling stage by using RAPD markers. Further RAPD markers were cost-effective and more convenient implication to pigeonpea. Earlier researchers used morphological or cytological approaches to analyze the interspecific hybrids of *Cajanus* [10, 14]. These cyto-morphological characters are controlled by environmental fluctuations and levels of polymorphism. Further, identification of the hybrids originating from *Cajanus platycarpus* and *Cajanus cajan* was carried out by using morphological markers and esterase isozyme patterns [8]. The diversity study of three cytoplasmic male sterile lines, *Cajanus cajan* genotypes and *Cajanus sericeus* genotypes were carried out using RFLP markers of mtDNA [21].

In this present study, a total of 20 RAPD primers of OPO series were screened, among them, 19 responded including 17 polymorphic primers, which generated 148 unequivocal scorable fragments, out of the 85 (57.43%) were polymorphic at the parent level. Among the parental polymorphic fragments, 41 and 44 fragments were unique to ICP26 and ICPW94 genotypes, respectively. The sizes of the amplified fragments were ranged from 225 to 3110bp. A total of 47 parental polymorphic fragments, including 21 female (ICP26) specific and 26 male (ICPW94) specific polymorphic fragments were inherited to F_1 hybrid. A total of 38 numbers of amplified fragments were present in both the parents but found in F_1 hybrids. Only seven primers (OPO01, OPO02, OPO05, OPO06, OPO07, OPO12 and OPO13) were able to show hybrid (25 markers) banding patterns betweenthe parents and the F_1 hybrid. Various authors used different series of RAPD makers for the genus *Cajanus* [9, 15, 16, 23]. RAPD markers were used for the identification of two pigeonpea CMS lines derived from the crosses between the wild *Cajanus scarabaeoides* and *Cajanus sericeus*, and the cultivated species of *Cajanus cajan* [23]. To find out the F_1 hybrids and their parents *Cajanus cajan* (ICP26) and *Cajanus scarabaeoides* (ICPW94), both the dominant markers RAPD and ISSR were used [9]. In this study, RAPD markers were used to determine maternity and paternity in the F_1 interspecific hybrid at the seedling stage, and this result could be moresupportive for more studies in chromosome mapping, marker-assisted breeding concerning wide hybridization.

Acknowledgments

Both the authors are very much thankful to the U. G. C., Govt. of India, New Delhi [F. No. 37-331/2009 (SR)] for financial help through the R&D project to the author (JP). Both authors are also obliged to Dr. H. D. Upadhyaya, Senior Scientist, ICRISAT for providing seeds required for the present study. The author (ARS) is especially thankful to the Chairman, the Director and the Vikash management groups for providing necessary facilities, moral support, and encouragement.

References

[1] Agbawa I. O., Patil, P., Datta, S., and Nadarajan, N. 2014, Comparative efficiency of SRAP, SSR and AFLP-RGA markers in resolving genetic diversity in pigeonpea (*Cajanus* sp.). *Indian Journal of Biotechnology*, 13, 486-495.

[2] Aruna, R., Rao, D. M., Reddy, L. J., Upadhyaya, H. D., and Sharma, H. C. 2005, Inheritance of trichomes and resistance to podborer (*Helicoverpa armigera*) and their association in interspecific crosses between cultivated pigeonpea (*Cajanus cajan*) and its wild relative *C. scarabaeoides*. *Euphytica,* 145, 247-257.

[3] Aruna, R., Rao, D. M., Sivaramakrishnan, S., Reddy, L. J., Bramel, B., and Upadhyaya, H. D. 2008, Efficiency of three DNA markers in revealing genetic variation among wild *Cajanus* species. *Plant Genetic Resources,* 7, 113-121.

[4] Datta, S., Kaashyap, M., and Kumar, S. 2010, Amplification of chickpea-specific SSR primers in *Cajanus* species and their validity in diversity analysis. *Plant Breeding*, 129, 334-340.

[5] Datta, S., Singh, P., Mahfooz, S., Patil, G. P., Chaudhary, A. K., Agbagwa, I. O., and Nadarajan, N. 2013, Novel genic microsatellite markers from *Cajanus scarabaeoides* and their comparative efficiency in revealing genetic diversity in pigeonpea. *Journal of Genetics*, 92, e24-e30.

[6] Ganapathy, K. N., Gnanesh, B. N., Gowda, M. B., Venkatesha, S. C., Gomashe, S. S., and Channamallikarjuna, V. 2011, AFLP analysis in pigeonpea (*Cajanus cajan* (L.) Millsp.) revealed close relationship of cultivated genotypes with some of its wild relatives. *Genetic Resources and Crop Evolution,* 58, 837-847.

[7] Lima-Brito, J., Carvalho, A., Martin, A., Heslop-Harrison, J. S., and Guedes-Pinto, H. 2006, Morphological, yield, cytological and molecular characterization of a breed wheat × tritorderm F_1 hybrid. *Journal of Genetics,* 85, 123-131.

[8] Mallikarjuna, N., and Moss, J. P. 1995, Production of hybrids between between *Cajanus platycarpus* and *Cajanus cajan. Euphytica*, 83, 43-46.

[9] Mishra, R. R., Sahu, A. R., Rath, S. C., Mishra, S. P., and Panigrahi, J. 2012, Cyto-morphological and molecular characterization of *Cajanus cajan* × *C. scarabaeoides* F_1 hybrid. *The Nucleus*, 55(1), 27-35.

[10] Mohanty, M., and Patnaik, S. N. 1989, Cytomorphological analysis of F_1 hybrids between *Cajanus cajan* (L.) Millsp. and *Atylosia cajanifolia* (Haines). *Cytologia*, 54, 121-128.

[11] Panguluri, S. K., Janaiah, J., Govil, J. N., Kumar, P. A., and Sharma, P. C. 2006, AFLP fingerprinting in pigeonpea (*Cajanus cajan* L. Millsp.) and its wild relatives. *Genetic Resources and Crop Evolution,* 53, 523-531.

[12] Panigrahi, J., Sahu, A. R., and Mishra, R. R. 2015, Development and applications of DNA markers in pigeonpea: The present status and Future prospects. In: *Applications of Molecular Markers in Plant Genome Analysis and Breeding*, edited by K. Taski-Ajdukovic and S. G. Pandalai Chapter-4 (Trivendrum, Kerala: Research Signpost), pp. 91-117.

[13] Patil, P. G., Datta, S., Agbawa, I. O., Singh, I. P., Soren, K. R., Das, A., Choudhary, A. K., and Chaturvedi, S. K. 2014, Using AFLP-RGA markers to assess genetic diversity among pigeonpea (*Cajanus cajan*) genotypes in relation to major diseases. *Acta Botanica Brasilica*, 28, 198-205.

[14] Pundir, R. P. S., and Singh, R. B. 1985, Biosystematic relationships among *Cajanus*, *Atylosia* and *Rhynchosia* species and evolution of pigeonpea. *Theoretical and Applied Genetics,* 69, 531-534.

[15] Ratnaparkhe, M. B., Gupta, V. S., Venmurthy, M. R., and Ranjekar, P. K. 1995, Genetic fingerprinting of pigeonpea (*Cajanus cajan* (L.) Millsp.) and its wild relatives using RAPD markers. *Theoretical and Applied Genetics*, 91, 893-898.

[16] Sahu, A. R., Mishra, R. R., and Panigrahi, J. 2016, Estimation of genetic diversity among 34 genotypes in genus *Cajanus* with contrasting host response to pod borer and its allied pests. *Journal of Crop Science and Biotechnology*, 19(1), 17-35.

[17] Sahu, A. R. 2015, *Construction of Genetic Map in Pigeonpea based on Inter-specific F_2 population using RAPD, SSR Markers and Simple Inherited Trait Loci (SITLs)*. PhD Dissertation, Sambalpur University, Odisha, India.

[18] Sahu, A. R., Mishra, R. R., Rath, S. C., and Panigrahi, J. 2015, Construction of interspecific molecular genetic map of pigeonpea using SCoT, RAPD, ISSR markers and SITLs. *The Nucleus*, 58(1), 23-31.

[19] Saxena, K. B., Singh, L., Reddy, M. V., Singh, U., Leteef, S. S., Sharma, S. B., and Remanandam, P. 1990, Intra-species variation in *Atylosia scaraebaeoides* (L.) Benth., a wild relative of Pigeonpea (*Cajanus cajan* (L.) Millsp.). *Euphytica,* 49, 185-191.

[20] Sharma, H. C., Pampapathy, G., and Reddy, L. J. 2003, Wild relatives of pigeonpea as a source of resistance to the pod fly (*Melanagromyza obtusa* Malloch) and pod wasp (*Tanaostigmodes cajaninae* La Salle). *Genetic Resources* and *Crop Evolution,* 50, 817-824.

[21] Sivaramakrishnan, S., Seetha, K., Rao, A. N., and Singh, L. 1997, RFLP analysis of cytoplasmic male sterile line of pigeonpea (*Cajanus cajan* L. Millsp.) developed by interspecific crosses. *Euphytica,* 93, 307-312.

[22] Smith, J. F., Burke, C. C., and Wangner, W. L. 1996, Interspecific hybridization in natural population of *Cyrtandra* (Gesneriaceae) on the Hawaian islands: evidence from RAPD markers. *Plant* Systematics and *Evolution,* 200, 61-77.

[23] Souframanien, J., Manjaya, J. G., Krishna, T. G., and Pawar S. E. 2003, Random amplified polymorphic DNA analyses of cytoplasmic male sterile and male fertile Pigeonpea (*Cajanus cajan* (L.) Millsp.). *Euphytica,* 129, 293-299.

[24] Varshney, R. K., Murali Mohan, S., Gaur, P. M., Gangarao, N. V. P. R., and Pandey, M. K., et al. (2013). Achievements and prospects of genomics-assisted breeding in three legume crops of the semi-arid tropics. *Biotechnology Advances,* 31, 1120-1134.

[25] Varshney, R. K., Penmetsa, R. V., Dutta, S., Kulwal, P. L., Saxena, R. K., Datta, S., et al. 2010, Pigeonpea genomics initiative (PGI): an international effort to improve crop productivity of pigeonpea (*Cajanus cajan* L.). *Molecular Breeding,* 26(3), 393-408.

[26] Verulkar, S. B., Singh, D. P., and Bhattacharya, A. K. 1997, Inheritance of resistance to pod fly and pod borer in the interspecific cross of pigeonpea. *Theoretical and Applied Genetics,* 95, 506-508.

Chapter 2

Strategies towards Household Food Processing for Lowering Antinutritional Factors in Chickpea (*Cicer arietinum* L.)

Anindita Roy*, Sanhita Ghosh and S. Kundagrami

Department of Genetics and Plant Breeding,
Institute of Agricultural Science,
University of Calcutta, Kolkata, India

Abstract

Chickpea (*Cicer arietinum L.*) are an old world grain legume cultivated for its unique compositions of nutrients in comparison to other plant foods. Sixty chickpea germplasms were collected from different sources among which twenty-five were selected as high yielding germplasms which assessed at Calcutta University's experimental farm, Baruipur during rabi season for four consecutive years. Chickpea seeds contained high level of antinutrients namely tannins, condensed tannins, phytic acid, trypsin inhibitors and α amylase inhibitors which hinder their nutritional and health benefits. Therefore an aim on reducing the anti-nutritional factors and improving the digestibility of in vitro protein and starch through microwave cooking and other household techniques such as soaking, boiling and pressure cooking of chickpeas were investigated. Based on this result, thermal treatments namely microwave cooking (9.10%-76.45%) greatly reduced the entire antinutritional factors while pressure cooking (30.38%-80.43%) was the best method in improving

* Corresponding Author's Email: aninditaroy999@gmail.com.

In: Proceedings of BIOSPECTRUM
Editors: Sanket J. Joshi, Susmita Mukherjee and Moupriya Nag
ISBN: 978

the digestibility of protein and starch through abolishing the anti-nutritional factors.

Keywords: chickpea, antinutritional factors, household techniques and digestibility of protein and starch

Introduction

Chickpeas (*Cicer arietinum* L.) holds a significant view in human diet owing to it's excellent source of minerals and vitamins and are preferred as a source of protein next only to milk especially to those living in developing countries (Singh et al., 2015). In spite of being the second most important pulse crop in the world and first in the Indian subcontinent (FAO) the productivity of chickpea in India is very low and unstable due to unavailability of healthy seeds. Therefore, breeding approaches should be made for identifying high yielding germplasms to enhance much-needed chickpea productivity. Another impediment in chickpea is that it also synthesizes certain biologically active substances as part of their protection to survive against adverse environmental conditions which devalue its nutritive quality like other legumes. These biologically active substances are known as antinutritional factors viz., trypsin inhibitor, alpha amylase inhibitors, phytic acid, tannin and condensed tannin which limits the potential utilization of nutrients (Wang et al., 2010). The activities of various enzymes present in the gut were inhibited by the presence of these types of antinutrients. Consequently, the digestion and absorption of protein and starch were hampered and interrupted the bioavailability of minerals. The improvement in the digestibility of chickpeas as affected by different food processing techniques which would help to transform raw grains into valuable products with maximum nutritional value to ensure the nutrient security of the population. Processing techniques at the household level found to be effective in improving the nutritive value of legumes. Thermal processing treatment viz., boiling, microwave and pressure cooking are widely accepted as an effective means of inactivating antinutrients of legume grains before consumption. Soaking has also been reported as an effective method in lowering the undesirable constituents in legume grains. Therefore objective of this experiment was to unravel the impacts of soaking and thermal treatments on reducing antinutritional factors and consequently enhancing the in vitro digestibility of protein and starch in selected high yielding chickpea germplasms.

Materials and Methods

Sample Collection

Sixty chickpea germplasms were obtained from NBPGR, Berhampore and some of them were local collection. The experiment was executed at Calcutta University's experimental farm, Baruipur, South 24 Parganas. The three replications in each year were done using randomized block design (RBD) during rabi season for four consecutive years from 2014 to 2018. Each germplasm was sown in a single row of 4 m length with a row to row spacing of 30 cm and 10 cm from plant to plant. Normal inter-culture operations were practiced throughout the growing periods. Harvesting of all germplasms were carried out manually at approximately 130-140 days after planting. Data was recorded of five best selected plants from each replication for eight traits.

Samples Preparation

Raw and dry chickpeas were taken as control.

Processing Methods

Chickpea seeds were soaked at room temperature for 9 hour in plain water at room temperature. The seed to solution ratio was 1: 3 (w/v). The soaking water was drained off, and rinsed twice in distilled water. For boiling effect, seeds were boiled in plain water at 100°C in the ratio of 1:10 (w/v) for 20 minutes. Chickpea seeds were taken in a bowl with plain water in a same ratio of 1:10, w/v and boiled in a microwave on high temperature for 15 min (until 50% of the seeds become soft). The pressure cooking was done at (1 kg/cm^2) using domestic pressure cooker with chickpea seeds and plain water ratio being 1:2. Cooking time was 5 minutes noted after the development of full pressure. All the treated seeds were oven dried at 55°C till constant weight and grinded until to pass through 5 mm sieve and preserved in deep freezer (-20°c) for further analysis.

Analysis of Antinutritional Factors

Estimation of trypsin inhibitor activity levels in chickpea were done following the method of Kakade et al., 1974. The method of Deshpande et al., 1982 was used to evaluate α - amylase inhibitory activity (AIA) in chickpea seedp. The phytic acid content of all treated chickpea seeds were measured by following to the method of Haug and Lantzsch 1983. Total tannins were extracted by the method described by Anonymous (2000). The method of Xu and Chang (2007) were followed to evaluate condensed tannin in chickpea seeds using the acidified vanillin reagent. IVPD of chickpea seeds were determined using the method described by Salgo′ et al., 1984. In vitro starch digestibility was assessed using the method given by Singh et al., 1982. These entire assays were done in triplicates with slight modifications.

Statistical Analysis

SPSS Program version 20 was used to perform all the statistical analysis. The results of mean performances of chickpea germplasms are expressed as mean ± SE. All the treatments data presented are presented as means± S.D. The statistically significant differences among all the samples at $P \leq 0.001$ was assessed by Duncan multiple range test (DMRT).

Results and Discussion

Mean Performance

High seed yield plant^{-1} is the main and foremost criteria in all breeding program. On the basis of yield performance over four years, twenty five high yielding chickpea germplasms were selected and presented in Table 1. Germplasms namely Digbijoy, BR78, S4, JG74 displayed considerable high values for different yield attributing traits.

Table 1. Mean performances of yield related traits of twenty five chickpea germplasms

Germplasms	Plant Height (cm)	No. of Branches Plant^{-1}	No. of Pods Plant^{-1}	Pod Length (cm)	No. of Seeds Pod^{-1}	100 Seed Weight (g)	Harvest Index	Seed Yield Plant^{-1} (g)
1. Digbijoy	75.13 ± 4.94^{a}	$6.59\pm0.47^{e-h}$	$85.17\pm8.51^{b-f}$	1.57 ± 0.03^{ab}	1.40 ± 0.21^{ab}	18.47 ± 0.09^{l}	51.81 ± 0.09^{gh}	22.02 ± 1.65^{ab}
2. Dcp92-3	63.82 ± 2.63^{ab}	6.98 ± 0.02^{h}	110.79 ± 6.81^{gh}	1.53 ± 0.02^{a}	1.42 ± 0.26^{ab}	12.62 ± 0.04^{f}	$50.08\pm0.04^{e-g}$	20.13 ± 0.83^{ab}
3. Avrodhi	62.79 ± 3.83^{ab}	$6.07\pm0.59^{b-h}$	$83.62\pm1.83^{b-d}$	2.02 ± 0.20^{f}	$1.68\pm0.09^{a-d}$	14.00 ± 0.24^{hi}	$48.67\pm0.24^{e-g}$	20.01 ± 2.18^{ab}
4. Radhey	70.95 ± 0.58^{ab}	$6.44\pm0.29^{d-h}$	$90.29\pm0.29^{c-f}$	$1.84\pm0.06^{d-f}$	$1.58\pm0.08^{a-d}$	11.50 ± 0.13^{d}	$46.69\pm0.13^{e-h}$	16.48 ± 1.85^{a}
5. PantG186	70.98 ± 1.73^{ab}	$6.62\pm0.31^{f-h}$	$78.10\pm1.60^{a-d}$	1.53 ± 0.14^{ef}	$1.46\pm0.04^{a-c}$	13.33 ± 0.20^{g}	$45.97\pm0.20^{d-g}$	16.04 ± 0.94^{a}
6. GNG469	65.03 ± 2.91^{ab}	$5.73\pm0.27^{a-f}$	$82.33\pm2.18^{b-e}$	1.86 ± 0.05^{ef}	$1.82\pm0.18^{a-d}$	11.23 ± 0.09^{cd}	$46.33\pm0.09^{e-g}$	17.14 ± 0.88^{ab}
7. Virat	68.33 ± 3.85^{ab}	$6.2\pm0.20^{c-h}$	$88.46\pm1.54^{b-f}$	$1.44\pm0.23^{a-f}$	$1.71\pm0.12^{a-d}$	12.24 ± 0.19^{ef}	$48.35\pm0.19^{e-h}$	18.44 ± 0.35^{ab}
8. B116	70.47 ± 3.18^{ab}	$6.06\pm0.06^{b-h}$	$78.52\pm3.52^{b-d}$	$1.69\pm0.08^{a-e}$	$1.83\pm0.08^{a-d}$	11.19 ± 0.03^{bd}	$40.28\pm0.03^{b-d}$	16.05 ± 0.14^{a}
9. BR 78	69.07 ± 6.61^{ab}	$6.00\pm0.02^{a-h}$	$87.72\pm2.64^{b-f}$	$1.82\pm0.07^{c-f}$	$1.70\pm0.06^{a-d}$	14.85 ± 0.22^{j}	$50.93\pm0.22^{e-g}$	23.20 ± 0.93^{b}
10. WBG29	71.70 ± 1.65^{ab}	$5.86\pm0.14^{a-g}$	$95.05\pm4.95^{b-f}$	1.95 ± 0.05^{ef}	2.32 ± 0.16^{e}	9.22 ± 0.11^{a}	$46.58\pm0.11^{e-g}$	18.03 ± 0.42^{ab}
11. Annigeri	74.18 ± 3.05^{ab}	$5.55\pm0.02^{a-d}$	$86.95\pm4.68^{b-f}$	2.00 ± 0.06^{f}	$1.79\pm0.29^{a-d}$	12.60 ± 0.08^{f}	$47.58\pm0.08^{e-h}$	19.02 ± 1.45^{ab}
12. RSG888	72.83 ± 2.59^{ab}	$5.61\pm0.11^{a-e}$	75.36 ± 2.63^{ab}	$1.79\pm0.10^{b-f}$	$1.93\pm0.07^{c-e}$	13.78 ± 0.17^{h}	$48.19\pm0.17^{e-h}$	19.72 ± 1.81^{ab}
13. HC5	62.76 ± 6.32^{ab}	6.81 ± 0.75^{gh}	98.88 ± 1.12^{fg}	$1.74\pm0.04^{a-f}$	$1.61\pm0.11^{a-d}$	11.92 ± 0.14^{e}	$50.00\pm0.14^{e-g}$	19.39 ± 2.65^{ab}
14. JG74	68.09 ± 4.08^{ab}	$6.29\pm0.29^{c-h}$	$90.94\pm9.06^{d-f}$	1.62 ± 0.05^{ef}	$1.64\pm0.14^{a-d}$	14.36 ± 0.25^{i}	$48.50\pm0.25^{e-g}$	20.47 ± 1.50^{ab}
15. L 550	71.96 ± 0.98^{ab}	$6.09\pm0.09^{b-h}$	64.87 ± 6.24^{a}	1.91 ± 0.06^{ef}	$1.81\pm0.02^{a-d}$	17.15 ± 0.15^{k}	$42.02\pm0.15^{b-e}$	18.02 ± 4.56^{ab}
16. IC268966	60.88 ± 4.44^{a}	$6.21\pm0.21^{c-h}$	$75.89\pm3.47^{a-c}$	$1.83\pm0.04^{c-f}$	$1.87\pm0.07^{b-e}$	12.18 ± 0.18^{ef}	33.87 ± 0.18^{a}	17.33 ± 1.66^{ab}
17. Rajesh	70.85 ± 0.60^{ab}	$5.47\pm0.47^{a-d}$	$79.92\pm2.44^{b-d}$	$1.79\pm0.03^{b-f}$	$1.65\pm0.35^{a-d}$	11.99 ± 0.01^{e}	36.27 ± 0.01^{ab}	16.16 ± 1.27^{ab}
18. BGM408	72.30 ± 1.87^{ab}	$5.9\pm0.10^{a-g}$	$95.75\pm4.25^{b-f}$	1.96 ± 0.04^{ef}	$1.78\pm0.05^{a-d}$	11.40 ± 0.04^{d}	39.85 ± 0.04^{bc}	19.34 ± 3.34^{ab}
19. M1	66.02 ± 3.77^{ab}	$5.79\pm0.21^{a-f}$	117.89 ± 4.27^{h}	$1.83\pm0.08^{c-f}$	1.34 ± 0.19^{a}	10.87 ± 0.35^{bc}	53.15 ± 0.07^{h}	18.79 ± 1.45^{ab}
20. M2	70.80 ± 4.17^{ab}	$5.43\pm0.43^{a-c}$	$90.79\pm0.61^{b-f}$	2.01 ± 0.04^{f}	1.97 ± 0.03^{de}	10.79 ± 0.05^{b}	$46.98\pm0.11^{e-h}$	17.02 ± 3.02^{ab}

Table 1. (Continued)

Germplasms	Plant Height (cm)	No. of Branches $Plant^{-1}$	No. of Pods $Plant^{-1}$	Pod Length (cm)	No. of Seeds Pod^{-1}	100 Seed Weight (g)	Harvest Index	Seed Yield $Plant^{-1}$ (g)
21. M3	64.15 ± 3.91^{ab}	$6.16\pm0.16^{b-h}$	$78.68\pm0.66^{b-d}$	$1.80\pm0.12^{c-f}$	$1.89\pm0.10^{b-e}$	11.29 ± 0.20^{cd}	$47.68\pm0.14^{e-h}$	17.16 ± 2.32^{ab}
22. M4	73.62 ± 6.35^{ab}	5.17 ± 0.17^{ab}	$85.56\pm6.06^{b-f}$	1.88 ± 0.10^{ef}	$1.90\pm0.05^{b-e}$	11.92 ± 0.03^{e}	$47.22\pm0.08^{e-h}$	17.22 ± 2.78^{ab}
23. S2	63.07 ± 7.59^{ab}	$5.93\pm0.07^{a-g}$	$82.57\pm3.80^{b-d}$	1.93 ± 0.07^{ef}	$1.94\pm0.06^{c-e}$	12.19 ± 0.10^{ef}	$44.93\pm0.19^{c-f}$	17.24 ± 0.52^{ab}
24. S3	66.43 ± 0.97^{ab}	$6.10\pm0.10^{b-h}$	94.78 ± 5.22^{ef}	$1.58\pm0.08^{a-d}$	$1.80\pm0.11^{a-d}$	13.03 ± 0.05^{g}	$50.18\pm0.01^{e-g}$	19.95 ± 1.98^{ab}
25. S4	72.68 ± 3.71^{ab}	5.05 ± 0.05^{a}	98.62 ± 1.38^{fg}	$1.56\pm0.06^{a-c}$	$1.54\pm0.04^{a-d}$	12.20 ± 0.06^{ef}	$50.09\pm0.10^{e-g}$	20.09 ± 3.30^{ab}

Note: Mean values of each column with different letters in the same row followed by different superscript letter (a-l) are significantly different at $P \leq 0.001$.

Antinutritional Factors

The twenty-five chickpea germplasms were subjected to soaking and various thermal treatments namely boiling, microwave and pressure cooking for reducing the level of different antinutritional factors (trypsin, phytic acid, tannin, condensed tannin and alpha amylase inhibitor). In this present study, only the top five chickpea germplasms exhibiting substantial reduction in antinutritional factors and significant improvement in in-vitro digestibility of protein and starch have been discussed and presented in table.

Trypsin Inhibitor Activity (TIA)

Raw chickpea had higher TIA content varying from 10.32 mg - 12.95mg/g. The influence of soaking followed by thermal treatment techniques are presented in Table 2. Although soaking had no significant effect on TIA contents, it reduced 4.75% - 9.85%. Similar observation was reported by Mubarak 2005 in mungbean seeds. The decrease in TIA content could be ascribed to the solubility of TIA into water during soaking. Heat treatments have been shown to be very effective in reducing TIA content (Alonso et al., 2000). Ordinarily, boiling reduced 62.55% - 63.38% TIA but didn't exhibit any significant effect in comparison with raw. While microwave cooking and pressure cooking were the most effective treatments in reducing TIA content significantly (71.04% - 76.45% and 75.66% - 80.43% respectively). Similar findings were reported by Mubarak, 2005 and Osman, 2007. Reduction of trypsin inhibitor during thermal processing might be due to interchange or destruction of the disulfide bonds in the protein structure of trypsin inhibitor (Alonso et al., 1998).

Table 2. Effects of household food processing on TIA (mg/g) contentin chickpea germpalsms

Variety	Raw	Soaking	Boiling	Microwave Cooking	Pressure Cooking
Digbijoy	$11.34{\pm}0.04^b$	$10.43{\pm}0.03^b$	$4.22{\pm}\ 0.03^b$	$3.17{\pm}0.02^b$	$2.76{\pm}0.02^b$
WBG 29	$11.83{\pm}0.03^c$	$10.94{\pm}0.04^c$	$4.43{\pm}0.03^c$	$3.34{\pm}0.04^c$	$2.83{\pm}0.03^c$
L-550	$10.32{\pm}0.03^a$	$9.83{\pm}0.04^a$	$3.65{\pm}0.05^a$	$2.43{\pm}0.03^a$	$2.02{\pm}0.03^a$
M1	$12.69{\pm}0.04^d$	$11.44{\pm}0.04^d$	$4.65{\pm}0.03^d$	$3.65{\pm}0.04^d$	$2.85{\pm}0.05^d$
S4	$12.95{\pm}0.03^e$	$11.89{\pm}0.75^e$	$4.85{\pm}0.03^e$	$3.75{\pm}0.04^d$	$2.95{\pm}0.05^d$

Note: Mean values of each column with different letters in the same row followed by different superscript letter (a-e) are significantly different at $P \leq 0.001$.

Phytic Acid (PA)

Raw chickpea contained high level of PA ranging from 7.85g - 11.03mg/g. The impacts of soaking followed by thermal treatments on PA content are presented in Table 3. During soaking, the decrease in PA content ranged from 9.49% - 13.69%. Similar results have been reported by other authors in legumes (Tajoddin et al., 2011 and Kaur et al., 2016). The decrease in PA content might be caused by soaking which induces activation of an endogenous enzyme phytase through diffusion of the products (Afify et al., 2011; Luo et al., 2009b). Better reduction in PA content has been observed during processing at elevated temperature and pressure. Boiling caused 12.60% to 22.68% reduction while microwave cooking 39.61% - 56.05% reduction in PA content. Pressure cooking treatment (43.91% -61.27%) significantly proved to be more effective than both boiling and microwave cooking (39.61%-56.05%). Vijayakumari et al., 2007 also reported a similar reduction in PA content after application of thermal treatment. According to El Maki et al., 2007 phytase activity at high temperature might degraded the inositol hexaphosphate to pentaphosphate or leads to a low molecular weight compound formation which helped in reduction of PA content.

Table 3. Effects of household food processing on PA (mg/g) content in chickpea germpalsms

Variety	Raw	Soaking	Boiling	Microwave Cooking	Pressure Cooking
Digbijoy	11.03 ± 0.03^{e}	9.95 ± 0.05^{e}	9.64 ± 0.04^{e}	6.64 ± 0.04^{e}	6.01 ± 0.05^{d}
WBG29	10.25 ± 0.05^{c}	9.13 ± 0.03^{c}	8.81 ± 0.01^{c}	6.19 ± 0.03^{c}	5.74 ± 0.05^{c}
L-550	7.85 ± 0.05^{a}	6.75 ± 0.05^{a}	6.07 ± 0.06^{a}	3.45 ± 0.05^{a}	3.04 ± 0.08^{a}
M1	9.69 ± 0.04^{b}	8.65 ± 0.05^{b}	8.02 ± 0.11^{b}	5.45 ± 0.05^{b}	5.04 ± 0.06^{b}
S4	10.75 ± 0.05^{d}	9.73 ± 0.03^{d}	9.05 ± 0.09^{d}	6.45 ± 0.05^{d}	6.03 ± 0.06^{d}

Note: Mean values of each column with different letters in the same row followed by different superscript letter (a-e) are significantly different at $P \leq 0.001$.

Tannin

The bitterness of chickpea is mainly due to the presence of tannin content which varied from 3.84mg – 4.92mg/g. All the treatments reduced tannin contents in comparison to raw. Effects of soaking and heat treatment on tannin contents in chickpea are presented in Table 4. However, soaking caused a decrease of 20.33% - 26.30% in tannin content which might be due to activation or *de novo* synthesis of enzymes through diffusion and hydrolysis of the water soluble antinutrients leaching out into the water (Vijayakumari et al., 2007). Some authors reported the reduction of tannins in other soaked seeds (Mubarak A.E. 2005 and Nithya et al., 2007). Consequently, boiling reduced the tannin contents (41.26% - 52.08%) but it had no significant effect. Treatments of pressure cooking had the maximum significant effect on lowering the tannin contents (52.24% -61.79%) whereas microwave cooking decreased 47.36%- 58.33% significantly. Habiba R.A. 2000 and Nithya et al., 2007 also observed decreased tannin contents in treated chickpea seeds. Moreover, this reduction of tannin content at high temperatures and pressure may be ascribed to the degradation of proteins or interaction of this compound with other components, such as proteins, to form insoluble complexes (Nithya et al., 2007).

Table 4. Effects of household food processing on Tannin (mg/g) content in chickpea germpalsms

Variety	Raw	Soaking	Boiling	Microwave Cooking	Pressure Cooking
Digbijoy	4.03 ± 0.03^{b}	3.04 ± 0.04^{b}	1.99 ± 0.04^{b}	1.72 ± 0.03^{a}	1.54 ± 0.03^{a}
WBG29	4.28 ± 0.02^{c}	3.25 ± 0.05^{c}	2.25 ± 0.03^{c}	1.93 ± 0.03^{b}	1.75 ± 0.04^{b}
L-550	3.84 ± 0.03^{a}	2.83 ± 0.04^{a}	1.84 ± 0.03^{a}	1.60 ± 0.05^{a}	1.50 ± 0.02^{a}
M1	4.63 ± 0.03^{d}	3.62 ± 0.03^{d}	2.63 ± 0.03^{d}	2.19 ± 0.04^{c}	2.04 ± 0.05^{c}
S4	4.92 ± 0.03^{e}	3.92 ± 0.03^{e}	2.89 ± 0.03^{e}	2.59 ± 0.04^{d}	2.35 ± 0.05^{d}

Note: Mean values of each column with different letters in the same row followed by different superscript letter (a-e) are significantly different at $P \leq 0.001$.

Condensed Tannin

Raw chickpea contained condensed tannin ranging from 492.20µg – 1573.05µg CAE/g. The influences of soaking and heat treatment on condensed tannins are presented in Table 5. Soaked caused a decrease of (6.35% - 23.80%) in condensed tannin content which may be attributed to the presence of high oligomeric polymers which are little soluble in water. During boiling, the decrease in condensed tannin content ranged from 33.46% - 37.01%. While microwave cooking and pressure cooking, reduced condensed tannin content ranging from 44.37% - 58.13% and 47.81% - 67.48% respectively. According to Kataria et al., 1989 during thermal treatment, it might undergo hydrolization resulting in the formation of insoluble low molecular weight polymers which helps in the reduction of condensed tannin content.

Table 5. Effects of domestic treatment on Condensed Tannin (µgCAE/g) content in chickpea germpalsms

Variety	Raw	Soaking	Boiling	Microwave Cooking	Pressure Cooking
Digbijoy	1573.05±0.05^{e}	1473.10±0.10^{e}	1016.07±0.08^{e}	875.06±0.07^{e}	802.05±0.05^{e}
WBG 29	1483.10±0.10^{c}	1372.05±0.05^{c}	976.10±0.10^{c}	775.09±0.08^{c}	706.07±0.08^{c}
L-550	492.20±0.26^{a}	310.05±0.05^{a}	206.06±0.06^{a}	176.08±0.08^{a}	160.04±0.06^{a}
M1	1506.08±0.08^{d}	1402.05±0.05^{d}	1002.09±0.08^{d}	801.09±0.08^{d}	786.06±0.06^{d}
S4	1450.05±0.05^{b}	1346.05±0.05^{b}	956.05±0.05^{b}	746.06±0.08^{b}	689.05±0.05^{b}

Note: Mean values of each column with different letters in the same row followed by different superscript letter (a-e) are significantly different at $P \leq 0.001$.

Alpha Amylase Inhibitor (AIA)

AIA content of chickpea varied from 48.65 - 52.05 U(Units)/g. Effects of soaking and heat treatment on AIA content in chickpea are presented in Table 6. Although soaking had no significant effect on AIA content, it reduced 9.39% - 11.12%. During boiling the decrease in AIA content varied from 38.39% - 45.39%, while microwave cooking reduced the AIA content 57.62% - 62.18%. Pressure cooking treatment (67.95% - 72.95%) significantly proved to be more effective than other cooking methods. Similar results of reduction had been found by Mulimani and Rudrappa 1993. Reduction of AIA content during thermal processing might be due to heat labile properties of AIA which becomes inactive at elevated temperature and pressure.

Table 6. Effects of household food processing on AIA (U/g) content in chickpea germpalsms

Variety	Raw	Soaking	Boiling	Microwave Cooking	Pressure Cooking
Digbijoy	50.65±0.04^{b}	45.25±0.05^{b}	30.25±0.07^{ab}	20.08±0.08^{b}	15.05±0.05^{b}
WBG 29	51.23±0.03c	46.14±0.05^{c}	31.24±0.04^{ab}	21.09±0.03^{c}	16.10±0.10^{c}
L-550	48.65±0.05^{a}	43.24±0.04^{a}	26.57±1.04^{a}	18.40±0.05^{a}	13.16±0.05^{a}
M1	52.05±0.05^{e}	47.16±0.05^{d}	32.07±0.08^{b}	22.06±0.06^{d}	16.68±0.21^{d}
S4	51.85±0.05^{d}	46.24±0.05^{c}	31.75±0.05^{b}	20.09±0.03^{b}	15.05±0.05^{b}

Note: Mean values of each column with different letters in the same row followed by different superscript letter (a-e) are significantly different at $P \leq 0.001$.

Table 7. Effects of household food processing on IVPD (%) content in chickpea germpalsms

Variety	Raw	Soaking	Boiling	Microwave Cooking	Pressure Cooking
Digbijoy	60.29±0.04^{a}	62.43±0.03^{a}	65.16±0.03^{a}	67.38±0.62^{a}	69.50±0.56^{a}
WBG 29	65.20±0.02^{b}	67.26±0.02^{b}	70.23±0.01^{b}	72.25±1.00^{b}	73.60±1.15^{b}
L-550	72.10±0.02^{e}	74.04±0.03^{e}	77.13±0.01^{e}	79.49±1.16^{d}	82.18±1.00^{d}
M1	67.86±0.06^{c}	69.86±0.02^{c}	72.90±0.02^{c}	75.22±1.16^{bc}	77.54±0.58^{c}
S4	69.76±0.04^{d}	71.87±0.03^{d}	74.76±0.01^{d}	76.11±0.57^{cd}	78.15±1.14^{c}

Note: Mean values of each column with different letters in the same row followed by different superscript letter (a-e) are significantly different at $P \leq 0.001$.

Table 8. Effects of household food processing on IVSD (%) content in chickpea germpalsms

Variety	Raw	Soaking	Boiling	Microwave Cooking	Pressure Cooking
Digbijoy	69.82±0.03^{a}	73.02±0.03^{a}	85.29±0.44^{a}	91.03±0.03^{a}	91.64±0.02^{a}
WBG 29	70.22±0.02^{b}	74.02±0.02^{b}	86.03±0.03^{ab}	92.73±0.03^{b}	92.85±0.03^{b}
L-550	71.02±0.03^{e}	75.02±0.03^{e}	87.23±0.03^{c}	94.24±0.02^{e}	94.72±0.03^{e}
M1	70.90±0.02^{d}	74.87±0.02^{d}	86.93±0.03^{c}	93.97±0.02^{d}	94.04±0.02^{d}
S4	70.56±0.02^{c}	74.78±0.02^{c}	86.57±0.02^{bc}	93.57±0.02^{c}	93.92±0.03^{c}

Note: Mean values of each column with different letters in the same row followed by different superscript letter (a-e) are significantly different at $P \leq 0.001$.

In Vitro Digestibility of Protein and Starch (IVPD and IVSD)

Raw chickpea contained IVPD and IVSD varying from 60.29%-72.10% and 69.82%-71.02% respectively. Effect of soaking and thermal treatments on IVPD and IVSD in chickpea germplasms are shown in Tables 7 and 8. Very little changes in IVPD and IVSD were found after soaking ranging from 2.69% - 3.55% and 4.58% - 5.98%. However, boiling improved the in vitro digestibility of protein varying from 6.98% - 8.08% consequently the effect of boiling on starch (22.16% - 22.82%) had significantly improved the digestibility. A significant improvement in both IVPD and IVSD were observed when raw seeds were subjected to microwave cooking and pressure cooking. The percent increase in IVPD during microwave cooking was 9.10% - 11.76%, whereas during pressure cooking it was increased 30.38% - 32.70%. The improvement in IVSD during pressure cooking was ranging from 31.25%-33.37% while improvement in microwave cooking was varying from 30.38%-32.70%. Luo and Xie 2013 observed that the structural changes of proteins such as globulin were promoted by thermal processing at high temperature and pressure. This structural alteration stimulated the penetration of protease enzyme by expanding the chain flexibility which improved the protein digestion. The rupture of starch granules of thermally treated chickpea germplasms, making the substrate more accessible which facilitated the amylolysis and improved the digestibility of starch.

Among the top five germplasms, S4 exhibited a significant improvement in in-vitro digestibility of protein and starch on attributing to the reduction of different antinutritional factors (Figure 1). Hence, S4 could be exploited as a promising germplasm in respect of food quality as well as yield in the global market in future.

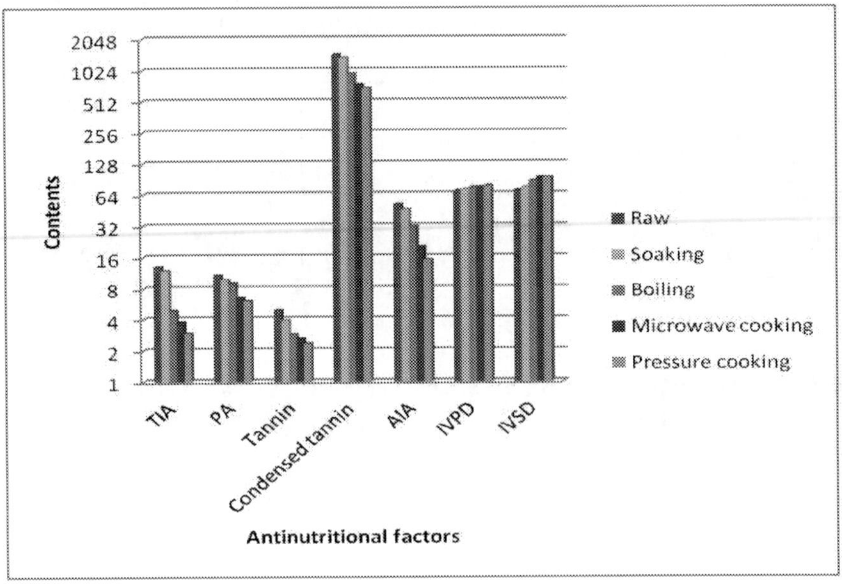

Figure 1. Effects of soaking and other thermal treatments on different antinutritional factors in chickpea germplasm S4.

Conclusion

Considering all the processing methods attempted in the present study, thermal treatments appeared to provide a maximum impact in improving the in vitro digestibility of protein and starch by reducing the antinutritional factors. However, it has been found that soaking played a little contribution in lowering the antinutrients. It is quite evident from the results that pressure cooking was the most effective treatment in reducing the various antinutritional factors of the chickpea germplasms. Conclusively, it can be stated that high productive chickpea germplasms with improved nutritional qualities using an inexpensive and convenient method has been identified from this study which could meet the nutritional requirements of the world population, particularly those of the developing nations.

References

Afify, A. E. M. M. R., El-Beltagi, H. S., Abd El-Salam, S. M., and Omran, A. A., 2011, Bioavailability of iron, zinc, phytate and phytase activity during soaking and germination of white sorghum varieties. *PLoS One*, 6 : 25512.

Alonso, R., Aguirre, A., and Marzo, F., 2000, Effects of extrusion and traditional processing methods on antinutrients and in vitro digestibility of protein and starch in faba and kidney beans. *Food Chemistry*, 68 : 159–165.

Alonso, R., Orue, E., Marzo, F., 1998, Effects of extrusion and conventional processing methods on protein and antinutritional factor contents in pea seeds. *Food Chem.*, 63: 505-512.

Anonymous, 2000, *Quantification of Tannins in Tree Foliage*, Vienna: Joint FAO/IAEA Division of Nucear Techniques in Food and Agriculture.

El-Maki, H. B., Abdel-Rahman S. M., Idris W. H., Hassan A. B., Babiker E. E., El-Tinay A. H., 2007, Content of antinutritional factors and HCl extractability of mineral from white bean (*Phaseolus vulgaris*). Cultivars: Influence of soaking and/or cooking. *Food Chemistry*, 100: 362–368.

Habiba, R. A., 2000, Changes in anti-nutrients, protein solubility, digestibility, and HCl-extractability of ash and phosphorus in vegetable peas as affected by cooking methods. *Food Chemistry*, 77, 187–192.

Haug, G., and Lantzsch, W., 1983, Methods for determination of phytate of cereal products. *Journal of the Science of Food and Agriculture*, 34, 1423–1424.

Kakade, M. L., Rackis, J. J., Mcgee, J. E., and Pushki, G., 1974, Determination of trypsin inhibitor activity of soy products: a collaborative analysis of an improved procedure, *Cereal Chemistry*, 51,376-382.

Kataria, A., Chauhan, B. M., and Punia, D, 1989, Antinutrients and protein digestibility (in vitro) of mung bean as a ected by domestic processing and cooking. *Food Chemistry*, 32 : 9-17.

Kaur S., Kaur S., Gupta A. K., and Kaur J., 2016, Physiochemical and nutritional attributes of raw and soaked seeds of chickpea (*Cicer arietinum* L.) genotypes. *Legume Research*, 39 (3): 359-369.

Luo, Y. W., and Xie, W. H., 2013, Effect of different processing methods on certain antinutritional factors and protein digestibility in green and white faba bean (Vicia faba L.). *CyTA – Journal of Food.*11 (1) : 43.49.

Luo, Y. W., Xie, W. H., Xie, C. Y., Li, Y., & Gu, Z. X., 2009b, Impact of soaking and phytase treatments on phytic acid, calcium, iron and zinc in faba bean fractions. *International Journal of Food Science and Technology*, 44, 2590–2597.

Mubarak, A. E., 2005, Nutritional composition and antinutritional factors of mung bean seeds (Phaseolus aureus) as affected by some home traditional processes. *Food Chemistry*, 89,489– 495.

Mulimani V. H. and Rudrappa G., 1993, Effect of heat treatment and germination on alpha Amylase inhibitor activity in chickpeas (Cicer arietinum L) 1993. *Plant Foods for Human Nutrition*, 46:133-137.

Nithya, K. S., Ramachandramurty, B., & Krishnamoorthy, V. V., 2007, Effect of processing Method on nutritional and antinutritional qualities of hybrid (COHCU-8) and traditional (CO7) pearl millet varieties in India. *Journal of BiologyScience*, 7, 643–647.

Osman, M. A., 2007, Effect of different processing methods, on nutrient composition Antinutrional factors, and in vitro protein digestibility of Dolichos lablab bean[Lablab purpuresus (L) Sweet]. *Pakistan Journal of Nutrition*, 6, 299–303.

Salgó, A., Granzler, K., Jecsai, J., 1984, Simple enzymatic methods for predication of plant protein digestibility. *Proc. Int. Assoc. Cereal Chem. Symp*. ed La´sztity R, Hidve´gi M Akade´miai Kiado´, Budapest, pp. 311–321.

Singh P. K., Shrivastava N., Sharma B., Bhagyawant S., 2015, Effect of domestic processes on chickpea seeds for antinutritional contents and their divergence. *American Journal of Food Science and Technology,* 3 (4): 111-117.

Singh U., Kherdekor M. S., Jambunathan R., 1982, Studies on desi and kabuli chickpea cultivars the levels of amylase inhibitors, level of oligosaccharides and in vitro starch digestibility. *J. Food Sci.*, 47: 510–512.

Tajoddin, M. D., Shinde, M. and Lalitha, J., 2011, In vivo reduction the phytic acid content of mungbean (Phaseolus aureus L) genotypes during germination. *American-Eurasian J. Agric. Environ. Sci.*, 10: 127 – 32.

Vijayakumari, K., Pugalenthi, M., & Vadivel, V., 2007, Effect of soaking and hydrothermal processing methods on the levels of antinutrients and in vitro protein digestibility of Bauhinia purpurea L. seeds. *Food Chemistry*, 103: 968–975.

Wang, N., Hatcher, D. W., Tyler, R. T., Toews, R. and Gawalko, E. J., 2010, Effect of cooking on the composition of beans (Phaseolus vulgaris L.) and chickpeas (Cicer arietinum L.). *Food Research International*, 43: 589–594.

Xu B. J., Chang S. K. C., 2007, A comparative study on phenolic profiles and antioxidant activities of legumes as affected by extraction solvents. *J. Food Sci.*, 72: 159–166.

Chapter 3

Nanobiofertilizers: A Boon in Agriculture

Aparna B. Gunjal*

Department of Microbiology,
Dr. D. Y. Patil, Arts, Commerce and Science College,
Pimpri, Pune, Maharashtra, India

Abstract

The use of chemical fertilizers to improve plant growth in agriculture has many disadvantages. The biological fertilizers are eco-friendly, economical and easy to use. The nanobiofertilizers are gaining wide attention in this aspect which is a boon in agriculture. The nanobiofertilizers promote the plant growth which is very helpful to the farmers in agriculture. This chapter focuses on nanobiofertilizers and their mechanisms. The chapter also describes the effect of nanobiofertilizers on plant growth.

Keywords: eco-friendly, economical, encapsulation, rhizosphere, sustainable

Introduction

Agriculture is very important aspect for farmers. To enhance the crop yield, farmers use chemical fertilizers. The chemical fertilizers are very toxic, costly and cause pollution affecting the plant growth and also the environment. The chemical fertilizers reach the soil and water bodies, thus causing soil and water

---

* Corresponding Author's Email: aparnavsi@yahoo.com.

In: Proceedings of BIOSPECTRUM
Editors: Sanket J. Joshi, Susmita Mukherjee and Moupriya Nag
ISBN: 978-1-68507-985-7

pollution. Therefore, it is very important to find some other solution to the use of chemical fertilizers.

The biological approach to improve the growth and yield of many plants and crops will be the best solution. The advantages of the biological approach are it is non-toxic, easy, and economical. This will also have sustainable approach. Therefore, use of biofertilizers is alternate to the use of chemicals. The fertilizers include phosphate and potassium solubilizing, nitrogen fixing, biofertilizers and nanofertilizers. In nanofertilizers, nanobiofertilizers are now gaining wide importance and it is the recent field of study in research to improve plant growth.

Nanotechnology has gained immense opportunities and novel applications in agriculture and biotechnology. And the emerging field in agriculture is nanobiofertilizers. Nanobiotechnology, together with nano-material and biofertilizer, will help to develop excellent nanobiofertilizer [1]. The nanobiofertilizers application will be important in optimizing photo-synthesis, mineral absorption, translocation and photosynthate accumulation [1]. The review here focuses on various aspects of nanobiofertilizers and their effect on plant growth.

Nanoencapsulation of Plant Growth Promoting Rhizobacteria

Nanotechnology has importance in agriculture with many applications. This includes bio-nano-encapsulation of beneficial bacteria which is important in agriculture scenarios. Nanofibers and other nanomaterials are eco-friendly and economical for inoculum development with PGPR, where the plant will benefit due to direct and indirect mechanisms of PGPR [2]. Nanofibers help to give protection to the bacteria and seeds from stress. The encapsulation of PGPR with nanofibers or nanomaterials is shown in Figure 1.

Nanobiofertilizers

Nanobiofertilizers are fertilizers made using wastes and other materials with the inclusion of indigenous microorganisms. These contain important nutrients in nano form providing enhanced efficiency and slow release. They need to be applied in very small amounts. The nanobiofertilizer is an eco-friendly approach for sustainable agriculture [1]. The great advantage of these nanobiofertilizers is that it is very economical as about a liter of nanobiofertilizer can be used in several areas of crops. The nanobiofertilizers include three types: macro and micro nanoparticles and nanobiofertilizers [3]. The three methods for application of nanobiofertilizers are foliar application [4], soaking the seeds before applying [5] and soil application [6].

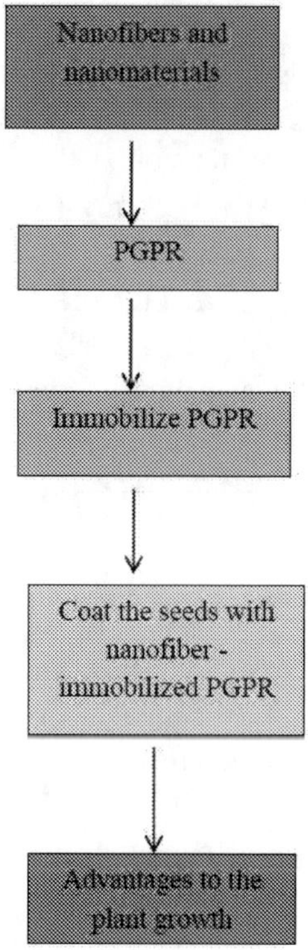

Figure 1. Encapsulation of nanofibers and nanomaterials with PGPR.

Nanobiofertilizers with PGPR

The application of chemical fertilizers by the farmers to the plants and crops is harmful, toxic as well as costly and also causes pollution which affects the environment [7]. Nanobiofertilizers will be an excellent alternate to the chemical fertilizers. Nanobiofertilizers prepared using PGPR will increase the nutrient supply to the plants, help to suppress the plant diseases, are safe when applied to the soil, and are economical and easy [2]. Nano bioformulations are excellent tools in environmental clean-up strategies [8]. The nano bioformulations enhance the adhesion of PGPR to the roots of the plants [9].

Mechanisms of Nanobiofertilizers

- Slow release of fertilizers will enhance plant growth effectively [10].
- Nanomaterial coatings on fertilizers hold the particles tightly from the plant due to large surface tension [11] and provide protection [12].
- Fertilizers with sulfur nano coating (≤100 nm) are beneficial slow release fertilizers [13].
- The nanobiofertilizer active ingredients are of the size 1-100 nm diameter which will increase the surface area, thereby increasing the absorption of minerals and other elements which will help in plant growth [14].

Nanofungicides

Fungi cause various diseases to many crops. This causes loss of many crops. Nanoparticles are effective as antifungal agents against pathogenic fungi. Antifungal activity of nanoparticles of zinc oxide (35-45 nm), silver (20-80 nm) and titanium dioxide (85-100 nm) have been reported against *Macrophomina phaseolina*, a pathogen of pulse and oilseed crops [15]. Similarly, silica nanoparticles are important antifungal agents against various phytopathogens [16].

Effect of Nanobiofertilizers on the Plant Growth

There is a report on nanobiofertilizers for nutrient use and sustainable productivity [3, 17, 18]. Also there is a report on use of nanofertilizers as foliar for *Cucurbita pepo* L. [19]. There is a report on significant increase due to nanobiofertilizers (combination of PGPR viz., *P. fluorescens*, *P. elgii* and *B. subtilis and gold nanoparticles*) [20]. Nano formulations of micronutrients have helped to enhance soil health and vigor index [21]. Nanofertilizer encapsulating nanosilica form binary films on the cell wall of fungi or bacteria and prevents infections. This will enhance plant resistance to different diseases [22]. Silicon-based nanobiofertilizers will improve vigor index and also root growth [23]. Study has been done on the use of nanobiofertilizers in crop production [24]. Study has been done on the effect of nanobiofertilizer on growth of spring wheat. Application of nanobiofertilizer (which contained *Azotobacter*, *Pseudomonas* along with Fe, Zn and Mn) has also shown to increase spike length, spike and seed number in square meters and seed weight. Also, there is study on nanobiofertilizer which will help in the growth of tomato and fenugreek [25]. Nanobiofertilizer promotes plant growth and nutritional quality due to diverse effects on soil and plant systems. Nanoencapsulation of organic nutrients using nanomaterials helps in slow release of nutrients [26]. The coating of biofertilizer by nanoparticles enhances nutrient stability. Nanobiofertilizer provides nutrients to the plants in a sustainable way [27]. The beneficial bacteria of nanobiofertilizer (e.g., PGPR or fungal inoculants) help the growth of crops by direct and indirect mechanisms viz., chelation of metals, solubilization of minerals, production of phytohormones, etc. [28, 29], thus enhancing the crop yield. Nanobiofertilizers have applications in horticulture [30, 31] as the chemical fertilizers are harmful, toxic and costly.

Conclusion

The nanobiofertilizers are eco-friendly, economical and promote the growth of many crops. The use of nanobiofertilizers is a biological approach which helps to maintain sustainability also. More research is needed in the area of nanobiofertilizers as it is gaining wide scope and attention.

References

[1] Kumar R. and Singh D. 2019. Nanobiofertilizer: An emerging eco-friendly approach for sustainable agriculture. *Proc Natl Acad Sci India Section B Biol. Sci* https://doi.org/10.1007/s40011-019-01133-6.

[2] Attia T. and Elsheery. 2020. Nanomaterials: Scope, applications, and challenges in agriculture and soil reclamation. In: Hayat S., Pichtel J., Faizan M. and Fariduddin

Q. (eds) *Sustainable Agriculture Reviews 41 Nanotechnology for Plant Growth and Development*. Springer, Switzerland.

[3] Chhipa H. 2017. Nanofertilizers and nanopesticides for agriculture. *Environ Chem Lett,* 15:15-22.

[4] Duhan J., Kumar R., Kumar N., Kaur P., Nehra K. and Surekha. 2017. Nanotechnology: The new perspective in precision agriculture. *Biotechnol Rep,* 15:11-23.

[5] Vakhrouchev A. and Golubchikov V. 2007. Numerical investigation of the dynamics of nanoparticle systems in biological processes of plant nutrition. *J Physics: Conf Series*, 61:31-35.

[6] Hafeez A., Razzaq A., Mahmood T. and Jhanzab H. 2015. Potential of copper nanoparticles to increase growth and yield of wheat. *J Nanosci Adv Tech,* 1:6-11.

[7] Caraglia M., Rosa G., Abbruzzese A., Leonetti C. 2011. Nanotechnologies: New opportunities for old drugs. The case of amino-bisphosphonates. *Nanomed Biotherapeu Dis,* 1:1-2.

[8] Huang S., Wang L., Liu L., Hou Y. and Li L. 2015. Nanotechnology in agriculture, livestock and aquaculture in China. A review. *Agron Sustain Develop,* 35:369-400.

[9] Mishra V. and Kumar A. 2009. Impact of metal nanoparticles on the plant growth promoting rhizobacteria. *Dig J Nanomater Biostruct,* 4:587-592.

[10] El-Ramady H., Abdalla N., Taha H., Alshaal T., El-Henawy A., Elmahrouk M., Bayoumi Y., Shalaby T., Al-Saeedi A., Amer M., Shehata S., Fári M., Domokos-Szabolcsy E., Sztrik A., Prokisch J., Pilon-Smits E., Pilon M., Selmar D., Haneklaus S. and Schnug E. 2018. Plant nano-nutrition: Perspectives and challenges. In: Gothandam K. et al., (ed) *Nanotechnology, Food Security and Water Treatment, Environmental Chemistry for a Sustainable World,* 11, Springer, Switzerland.

[11] Subramanian K., Manikandan A., Thirunavukkarasu M. and Rahale C. 2015. Nanofertilizers for balanced crop nutrition. In: Ra M. et al., (eds) *Nanotechnologies in Food and Agriculture*. Springer, Switzerland. pp. 69-80.

[12] Brady N. and Weil R. 1999. *The Nature and Properties of Soils.* New Jersey, USA, Prentice Hall. pp. 415-473.

[13] Subramanian K. and Thirunavukkarasu M. 2017. Nanofertilizers and nutrient transformations in soil. In: Ghorbanpour M., Manika K. and Varma A. (eds) *Nanoscience and Plant-Soil Systems. Soil Biology Series*, 48. Springer, Switzerland. pp. 305-319.

[14] Manjunatha S., Biradar D. and Aladakatti Y. 2016. Nanotechnology and its applications in agriculture: A review. *J Farm Sci,* 29:1-13.

[15] Shyla K., Natarajan N. and Nakkeeran S. 2014. Antifungal activity of zinc oxide, silver and titanium dioxide nanoparticles against *Macrophomina phaseolina. J Mycol Plant Pathol*, 44:268-273.

[16] Suriyaprabha R., Karunakaran G., Kavitha K., Yuvakkumar R., Rajendran V. and Kannan N. 2014. Application of silica nanoparticles in maize to enhance fungal resistance. *IET Nanobiotechnol,* 8:133-137.

[17] Kalia A. and Kaur H. 2019. Nanobiofertilizers: Harnessing dual benefits of nano-nutrient and biofertilizers for enhanced nutrient use efficiency and sustainable

productivity. In: Pudake R., Chauhan N. and Kole C. (eds) *Nanoscience for Sustainable Agriculture*. Springer, Cham.

[18] Singh A. and Prasad S. 2016. Nanotechnology and its role in agroecosystem: A strategic perspective. *Int J Environ Sci Technol*, 14:2277-2300.

[19] Hassan A., Salama D., Abd El-Aziz M. and Abd Elwahed M. 2019. Green synthesis of nano fertilizers and their application as a foliar for *Cucurbita pepo* L. *J Nanomaterials*, 2019:1-11.

[20] Shukla S., Kumar R., Mishra R., Pandey A., Pathak A., Zaidi M., Srivastava S. and Dikshit A. 2015. Prediction and validation of gold nanoparticles (GNPs) on plant growth promoting rhizobacteria (PGPR): A step toward development of nanobiofertilizers. *Nano Rev*, 4:439-448.

[21] Peteu S., Oancea F., Sicuia O., Constantinescu F. and Dinu S. 2010. Responsive polymers for crop protection. *Polymer,* 2: 229-251.

[22] Wang L., Wang Y., Li M., Fan M., Zhang F., Wu X., Yang W. and Li T. 2002. Synthesis of ordered biosilica materials. *Chin J Chem,* 20:107-110.

[23] Hutasoit S., Suada I. and Susrama I. 2013. Antifungal activity test extract some type of marine life link to *Aspergillus flavus* and *Penicillium* sp. *E J Trop Agroecotechnol,* 2:27-38.

[24] Danish M. and Hussain T. 2019. Nanobiofertilizers in crop production. In: Panpatte D. and Jhala Y. (eds) *Nanotechnology for Agriculture: Crop Production and Protection*. Springer, Singapore.

[25] Hussein H., Shaarawy H., Hussein N. and Hawash S. 2019. Preparation of nanofertilizer blend from banana peels. *Bull Nat Res Centre*, 43:1-9.

[26] Qureshi A., Singh D. and Dwivedi S. 2018. Nanofertilizers: A novel way for enhancing nutrient use efficiency and crop productivity. *Int J Curr Microbiol App Sci*, 7:3325-3335.

[27] Vejan P., Abdullah R., Khadiran T., Ismail S. and Boyce A. 2016. Role of plant growth promoting rhizobacteria in agricultural sustainability - A review. *Molecules*, 21:573.

[28] Ahemad M. and Kibret M. 2014. Mechanisms and applications of plant growth promoting rhizobacteria: Current perspective. *J King Saud Uni - Sci*, 26:1-20.

[29] Mala R., Selvaraj R., Sundaram V., Shanmuga Rajan R. and Gurusamy U. 2017. Evaluation of nanostructured slow release fertilizer on the soil fertility, yield and nutritional profile of *Vigna radiata*. *Recent patents on Nanotechnol,* 11:50-62.

[30] Chinnamuthu C. and Boopathi P. 2009. Nanotechnology and agroecosystem. *Madras Agric J*, 96:17-31.

[31] Shaviv A. 2000. Advances in controlled release of fertilizers. *Adv Agron J,* 71:1-49.

Chapter 4

A Study of Different Types of Carcinogenic Food Components, Food Colours and Preservatives Used in Food Items of Regular Use: A Review

Dipro Mukherjee[2], Sayantani Garai[2], Bandita Dutta[1], Rina Rani Ray[1], Sanchari Das[2], Vijay Jagdish Upadhye[3], Dibyajit Lahiri[2,*] and Moupriya Nag[2,†]

[1]Department of Biotechnology,
Maulana Abul Kalam Azad University of Technology,
West Bengal, India
[2]Department of Biotechnology,
University of Engineering and Management,
Kolkata, India
[3]Center of Research for Development (CR4D),
Parul Institute of Applied Sciences (PIAS),
Vadodara, Gujarat, India

Abstract

Food is one of the most important sources of life. In the present twenty-first century, the quantity of food produced is enough to serve all humanity but still, a major percentage of people starve and die from

---

* Corresponding Author's Email: dibyajit.lahiri@uem.edu.in.
† Corresponding Author's Email: moupriya.nag@uem.edu.in.

In: Proceedings of BIOSPECTRUM
Editors: Sanket J. Joshi, Susmita Mukherjee and Moupriya Nag
ISBN: 978-1-68507-985-7

hunger. The world suffers from the problems of undernutrition and overnutrition and most importantly, people suffer from cancer. Cancer is one of the leading causes of mortality just after cardiovascular diseases. In this study, we aim to summarize the present relation between processed food items, canned food, fast food, food preservatives, and food colors which comprise carcinogenic agents and cause cancer. The varied ways and vital ingredients present within these food items that are responsible for causing different types of cancer within the human body are thoroughly discussed. This study also lights on some famous food colours used often in India and other Asian countries that trigger the early onset of cancer.

Keywords: epigenetics, salted fish, aspartame, cancer, artificial colour, nitrosamines

Introduction

Cancer is the deadliest disease in this world presently that has affected billions of people for a large period [1]. The onset of cancer can take place in almost any part of our body independent of the age of any person. Cancer means uncontrolled cell division leading to the formation of a tumor. Various types of cancer exist which can vary from lymphoma, carcinoma, sarcoma to organ-specific cancers like breast cancer, bladder cancer, urinary cancer, liver cancer, prostate cancer, etc. It starts from a single cell and occurs due to mutation in the homeostasis mechanism or the presence of substances known as carcinogens in our body [2]. Carcinogens create a mutation in the gene or they can cause modification and repairing of the nucleic acids. But in most cases, the repair is not done correctly leading to the onset of the primary stage of cancer. Figure 1 shows various stages of cancer during the disease progression. The presence of carcinogenic promoters can be traced in this stage and monitored to track changes in a mutagenic cell to produce an abnormally functioning cell [3] (Figure 2).

The initial progress of the disease is affected both by environmental factors as well as dietary consumptions. Considering the latter case, certain food components present naturally or artificially are carcinogenic and require immediate attention for the prevention of cancer. This may include processed food items, canned food, fast food, food preservatives and food colours.

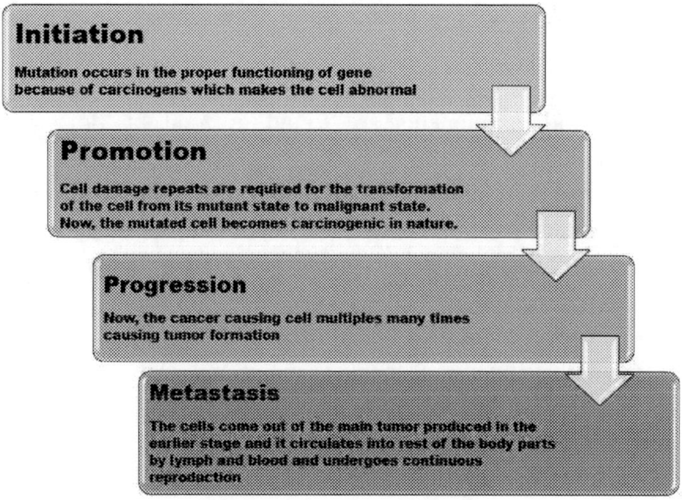

Figure 1. Different stages of cancer.

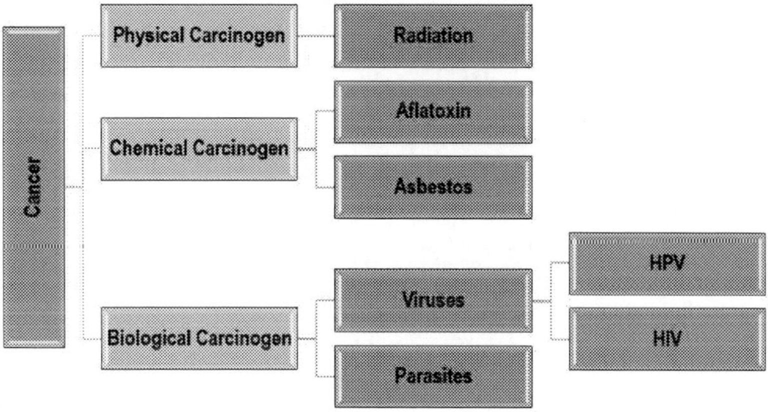

Figure 2. Different types of artificial carcinogens.

Some Natural Food Toxins

Food toxins are certain components that are naturally present or artificially introduced and proven to be unfit for human consumption and in worst cases they might be carcinogenic too [4].

These carcinogenic components are primarily subdivided into four types as follows:

1. The natural components are found in food by default and are unavoidable.
2. The components which initially were not found to be carcinogenic however may turn into a carcinogenic form due to some fungal or microbial activities.
3. The components may themselves be anthropogenic.
4. The components are artificially introduced in the food and are proved to be anthropogenic [5]. They are many examples of naturally occurring carcinogenic compounds (Figure 3). Aflatoxins are a group of toxic metabolites produced by certain species of fungi, especially Aspergillus flavus and Aspergillus parasiticus [6]. Once they enter the food chain, they affect all the levels of the food web. Toxins from Fusarium moniliforme, Fumonisins, Fumonsin B1 that is produced by a species of fungi known as Fusarium verticilliodes, is the sequence of toxic components that are responsible for causing leuko-encephalomalakia and porcine pulmonary especially to farm animals [10]. Cancer can also be caused by all types of alcoholic beverages that comprise ethanol, harmful for human consumption. They are the major cause of liver cirrhosis. Ligands of the aryl hydrocarbon receptor (AR) such as 2,3,7,8-Tetracholordibenzo-p-dioxin are known for their vigorousness towards high AR binding affinity and their resistance to either environmental or biological degradation [7]. It is majorly responsible for producing tumors in the liver, thyroid, the upper aerodigestive tract, and skin. Humans are often exposed to acetaldehyde in the air, from several foods and it is often the product of different geological, industrial, and biological processes. Exposure to Polycyclic aromatic hydrocarbons (PAH) including Benzo[a]-pyrene which is a typical member of the polycyclic aromatic hydrocarbon family, which also includes benzo[a]anthracene, benzo[b]fluoranthene, and other similar compounds that have the same toxic profiles through various routes involving food, therapeutic use of coal tar, occupation exposure to PAHs leading to tumors in various body parts [8]. N-Nitroso dimethylamine and some other nitrosamines are often formed during rubber processing, smoked and preserved food, in alcoholic beverages, drugs, tobacco smoke, and are

also present in metalworking fluids. Hot Mate is an herbal infusion of the plant Ilex paraguariensis, whose leaves are found to be a carcinogenic agent that may cause cancer in the oral, oropharyngeal, head and neck, and oesophageal zone. Bracken Fern is a member of the family Dennstaedtiaceae and can cause gastric and oesophageal cancer when directly consumed [9]. Dichlorodiphenyltrichloroethane (DDT) is the most common and well-known pesticide. Once it enters the food chain it accumulates in the form of fat as it is lipophilic in the higher consumption of the food web.

Humans are exposed to Ochratoxin A, produced by molds such as Aspergillus, Penicillium through several routes such as foods, animal products, and processed food including beer. It has a carcinogenic effect and mainly targets the liver and kidneys [11]. The temperature and form of cooking are responsible for changing the production of PhIP, an arylamine, and two related heterocyclic amines, namely 2-amino-3,4-dimcthylimidazo[4,5-f] quinolone (MeIQ) and 2-amino-3,8-dimethylimidazo[4,5-f] quinoxaline (MeIQx) which are considered to be mutagenic and cause cancer majorly in animals and possibly to humans, especially in intestine, prostate, lymphomas, and other sites. High temperature may also lead to the production of Acrylamide [12]. Studies have shown several pieces of evidence that prove coffee to be carcinogenic to humans and it increases the risk of colon and bladder cancer. The vegetable during pickling may gradually produce anthropogenic and mutagenic substances that increase the risk of oesophageal cancer in consumers [13]. Moreover, this process may also involve the growth of molds which probably may not fit for human consumption [14].

Several components in our regular food may prove carcinogenic to humans like spices. These have always provided benefits to human health but can lead to disease if ingested in excess. For example, the component named coumarin in cinnamon is toxic when consumed in excess. Another example is sugar [15]. Every food contains sugar in it for enhancing taste. However, the high fructose level in it is responsible for causing serious health issues including obesity, diabetes, fatty liver. Excess sugar intake can lead to tumour growth leading to cancer in breast and colon areas. Fish is rich in omega 3 fatty acid which is major food for brain. It improves the quality of hair and skin. It is also good for the heart [16]. However, certain varieties of fish majorly marine fish contain a high level of known toxic, mercury which enters the food chain through consumption and undergoes bioaccumulation. Several plant species that humans consume as food contain cyanogenic glycosides which result in cyanide intoxication if exceeds the safe intake level may lead to rapid respiration, drop in blood pressure, dizziness, headache, stomach pains, vomiting, diarrhoea, mental confusion, terminal coma, and death [17]. When a plant is exposed to physical damage furocoumarins, often described as stress toxins are released in response to stress. They are responsible for causing gastrointestinal problems and severe skin reactions under sunlight (UVA exposure) [18]. Lectins are the toxins majorly found in several types of beans that can cause severe stomach aches, vomiting and diarrhea. Mycotoxin, a natural toxic compound in moulds is found to be grown on numerous foodstuffs and can cause long-term effects on health which include cancers and immune deficiency [19]. Solanines and chaconine, are glycoalkaloids found in all Solanaceae plants. These are the natural toxins that are produced by the plants in response to stresses like bruising, UV light, microorganisms, and attacks from insect pests and herbivores [20]. Muscimol and muscarine are toxins found in wild mushrooms which can cause vomiting, diarrhoea, confusion, visual disturbances, salivation, and hallucinations. Many plant species belonging to the families Boraginaceae, Asteraceae, and Fabaceae often fall under the category of weeds and produce toxins named as Pyrrolizidine Alkaloids (PAs) [21]. They can grow in fields and contaminate food crops and can severely give rise to health-related problems leading to cancer and DNA damaging. Cows nowadays are injected with Recombinant Bovine Growth Hormone (Rbgh) for boosting milk production and it contains high levels of IGF-1 which cause various types of cancer [22]. Rich in omega-6 fats Refined vegetable oil can cause heart disease and cancer.

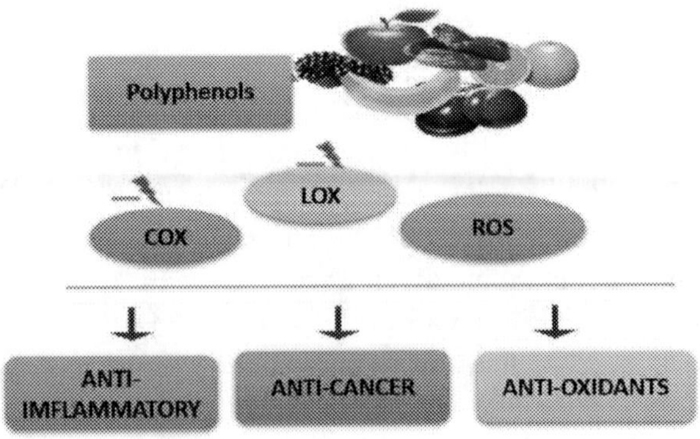

Figure 3. Effects of diet habits on cancer development.

Chewing gums contains camauba wax which can cause cancer and tumours. The artificial sweeteners and food colouring which are added in the food for enhancing taste have also been proved carcinogenic to humans [23]. Different components in packaging food may also have carcinogenic effects. Many foods and beverages are contained in plastic containers which contain bisphenol-A (BPA) [24] that may leach into our food and may lead to health issues (Figure 4). The ink in newspapers which are commonly used for wrapping hot oily food is carcinogenic and toxic. It may enter into our food and cause adverse health effects. Aluminium which is used for wrapping and decoration may gradually leech into our food and may cause severe health problems [25]. There are several food additives that have been proved to have adverse health effects such as Sodium nitrate, Sulphites, Azodicarbonamide, Potassium bromates, Propyl gallate, BHA/BHT, Propylene glycol, Butane, Monosodium glutamate (MSG), Disodium inosinate, Disodium guanylate, Enriched flour, Sodium benzoate, Propyl gallate, Olestra, Carrageenan, Polysorbate 60, Chlorine dioxide and Paraben [26]. To protect ourselves from all these carcinogenic toxic components we have to maintain several steps such as replacement of artificial colours with natural colours often obtained from beetroots, raspberries, red cabbages, leaves, stems, apricots, carrots and tomatoes, preventing any further use of damaged, pungent taste or smell discolored, mouldy food [27, 28].

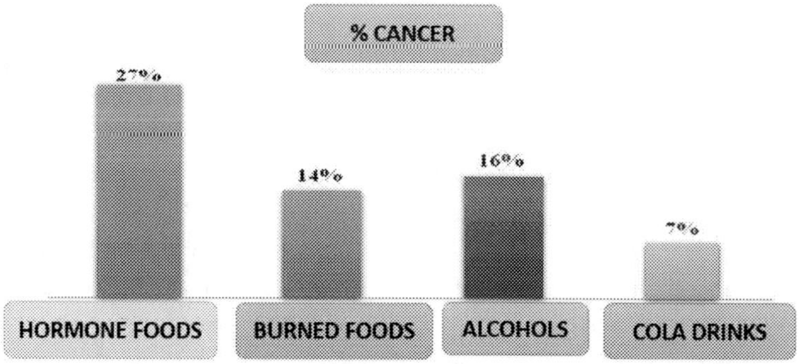

Figure 4. Percentage of commercial food products on development of cancer.

Risk Factor

Habits like consumption of alcohol, sexual intercourse with multiple sex partners, smoking, excessive consumption of red meat, processed food and drug intake are one of the most dangerous factors that can cause cancer [29]. Thus, procuring good habits like stopping consumption of tobacco and alcohol, preventing obesity and following a healthy diet, undergoing regular physical activities, preventing HIV infection, preventing household smoke and air pollution in urban areas will help prevent more than 30% of all cancer death across the world [30, 31, 32].

Epigenetics

The term epigenetics means a study of physiological and cellular characters which are not created by DNA sequence changes but which can still be inherited by the daughter cells [33]. Specifically, Epigenetics is the study of long-term alterations that are stable inside the cell's transcriptional potential [34]. It includes events like modification of covalent histones specifically the methylation and histone acetylation, mechanisms which are in relation with non-coding RNAs and methylation of the DNA [35]. By the study of Epigenetics, it has been seen that every type of cancer is caused by chemical changes which is caused either by the food habits or the environment factors [36]. Cancer caused due to these factors doesn't make it a genetic disorder but makes it gene function alteration. That's why nowadays, many research works are conducted for finding the treatment against alternation in the function of a gene [37]. This however, doesn't mutate the entire gene. We all know that gene expression is changed by Epigenetic phenomenon [38]. For an instance, tumour silencing suppressor gene usually has a regulatory activity upon cell growth and cell death, by causing not only a cell mutation but also an epigenetic change. So, these modifying characteristics has a significant role in occurrence of any type of diseases especially cancer [39]. For example, through some of the recent conducted studies, we come to know that consumption of too much acrylamide can cause DNA hypomethylation which again can result in permissive transcriptional activity at the centromere [40, 41] (Table 3).

Again, over consumption of Aflatoxin M1 can cause accumulation of tiny sized satellite transcripts which can impair centromeric structure and its activity [42]. Overconsumption of Nitrosamines again can result in hypomethylation of the repetitive elements at the sub-telomeric areas [43]. This can result in increased rate of transcription of the telomeres leading to an abnormal condition [44, 45].

Components Used in Food Having Carcinogenic Properties

Some compounds are seen to be used in different food items that starts a reaction and gives rise to some modifications in epigenetics that can cause cancer (Tables 1 and 2). These components are used in our regular food consumption [46] (Figure 5).

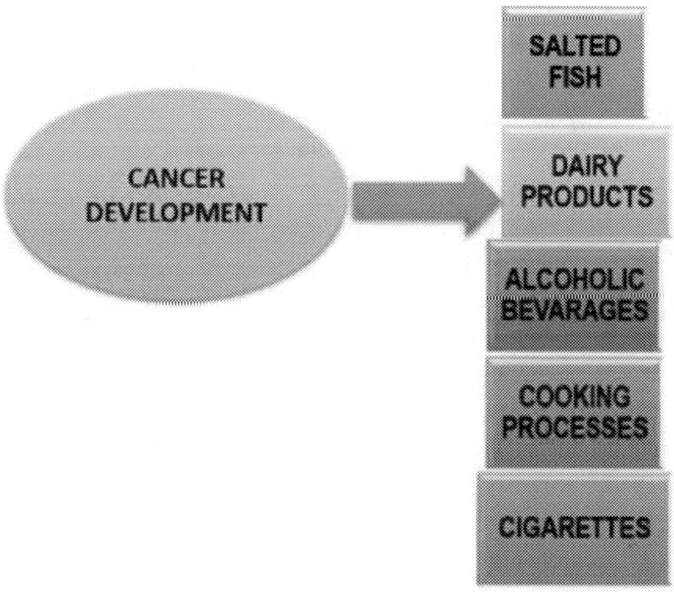

Figure 5. Food Components helping in cancer progression.

Table 1. Commercial food products which are carcinogenic, probably carcinogenic and non-carcinogenic in nature

	IARC Group 1: Carcinogenic to Humans	
SL No.	Compound Names	References
1.	Benzene	[36, 37]
2.	Salted fish (Chinese style)	[39]
3.	Arsenic and arsenic compounds, arsenic in drinking water	[43, 47]
4.	Benzene Aflatoxins (naturally occurring mixtures)	[36]
5.	2,3,7,8-Tetrachlorodibenzo-para-dioxin	[35, 65]
6.	Cadmium and cadmium compounds	[31, 32]
7.	Alcoholic beverages, ethanol in alcoholic beverages	[33]
8.	Benzo[a]pyrene	[43, 44]
	IARC Group 2A: Probably Carcinogenic to Humans	[47]
9.	Hot mate	[65]
10.	N-Nitrosodiethylamine, N-nitrosodimethylamine	[76, 77]
11.	High-temperature frying,	[73, 75]
12.	Ethyl carbamate (urethane)	[21, 11]
13.	Nitrate or nitrite (ingested under conditions that result in endogenous nitrosation)	[13, 14]
14.	Acrylamide	[15]
15.	Some pesticides (e.g., captafol)	[34]
	3. IARC Group 2B: Possibly Carcinogenic to Humans	[43, 44]
16.	Fumonisin B_1	[63, 69]
17.	Safrole	[51]
18.	Aflatoxin M_1	[53]
19.	Some heterocyclic amines	[11, 17]
	3. IARC Group 2B: Possibly Carcinogenic to Humans	[43, 44]
20.	Coffee (urinary bladder)	[19, 27]
21.	Acetaldehyde	[29]
22.	Some polycyclic aromatic hydrocarbons	[31, 33]
23.	Ochratoxin A	[19, 21]
24.	Nickel, metallic	[43]
25.	Furan	[49]
26.	Pickled vegetables	[81, 83]
27.	Acetamide	[89, 87]
28.	Some pesticides	[88]
29.	Lead	[12]
30.	Caffeic acid	[21]

Table 2. Carcinogenic or probably carcinogenic compounds present in different commercial food products

Sl no.	Component	Food
1.	Acrylamide	French fries, potato chips, cornflakes
2.	Neu5GC	Red meat
3.	Sodium nitrite i.e., nitrosamines	Processed meat
4.	PFOA (perfluorooctanoicacid)	Microwave popcorn
5.	Nitrosamines	Salted fish
6.	Aspartame	Soft drinks (diet)
7.	Saturated fats	Butter/oil/milk

Table 3. Carcinogenic compounds present in different food products and the diseases caused by them

Sl No.	Disorders they may cause	Food constituent
1.	Diabetes	Acrylamide from French fries
2.	Phenylketonuria, Alzheimer (yet to be proved)	Aspartame from canned drinks
3.	Cholesterol	Saturated Fats from butter and oils
4.	Diabetes and Alzheimer	Nitrosamines from pickles and red meat
5.	Polymer Fume Fever – which is a type of flu which occurs due to inhalation of PFOA fumes. It is not lethal to animals.	PFOA from non-stick ware coating

Acrylamide

The consumption surveys of FDA reports that eight food items namely, breakfast cereal, potato chips, French fries, soft bread, toast, brewed coffee and cookies shows that the mentioned food items can be carcinogenic to human health and they can also induce a certain amount of toxicity [47]. Among these food items, potato chips and French fries has the most consumption rate. This is because compared to the other type of foods, these have an easy preparation process, they have multiple categories and flavours for attracting children and they are comparatively cheap [48]. Potatoes being the main ingredient of these French fries [49], it possesses a handsome amount of lipids and carbohydrates, along with some amino acids such as asparagine.

Acrylamide Formation Process

Preparation of French fries and potato chips are done by deep frying of potatoes at temperatures as high as 120 degrees Celsius [50]. High temperature during frying or cooking or baking converts carbohydrate to form acrylamide along with asparagine through a reaction known as Maillard reaction [51].

Carcinogenicity Induced by Acrylamide

It has been proven through many studies that acrylamide has a significant amount of carcinogenicity in Laboratory animals [52]. But the exact effect in human is still not effectively proven. However, through many experiments conducted in different universities throughout the world proves that a relationship can be established among increased acrylamide consumption level on one hand and increasing chances of cancer on the other [53]. Again, some of the research works also confirms that there is much relevant evidence that can be found out about relationship between higher level of acrylamide consumption and increased chances of having colon cancer in females [54].

Again, another study was performed to determine the relationship between increased amounts of acrylamide and higher chances of breast cancer in females [55]. This experiment provided much concrete evidence towards the relationship between the two [56].

Inhibitors of Acrylamide

Many experiments have been done previously to prevent the accumulation of higher levels of acrylamide in its manufacture procedure [57]. Many experiments have also conveyed that the usage of pectin and alginic acid might undergo inhibition of the production of acrylamide [58, 59]. The time span of soaking of potatoes also has a significant function in the inhibition of acrylamide production.

Nitrosamines

Nitrosamines is found in the form of NOCs or the N-nitroso-compounds [60]. This compound gets in our system through tobacco smoking, diet, drinking water and work place. Nitrosamines can also be present in the food items we consume or that is formed because of the multiple processes or reactions [61, 62]. Figure 6 depicts the production process of Nitrosamine inside Stomach.

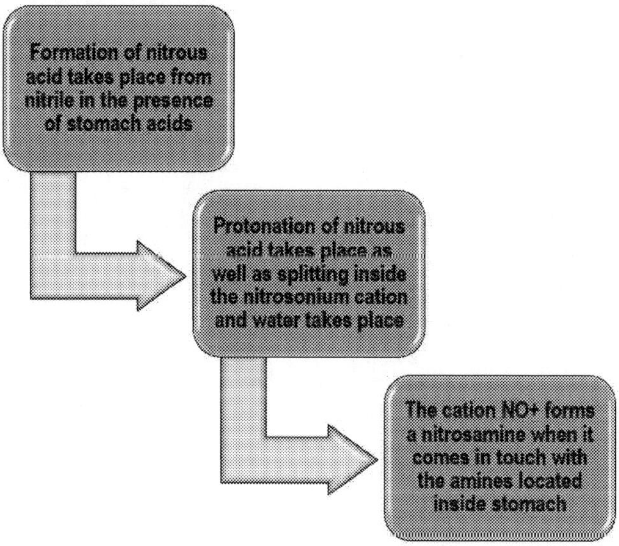

Figure 6. Production process of nitrosamine inside stomach.

Formation of Nitrosamines

Nitrosamines are produced in acid rich surroundings in the stomach as a result of a reaction among the amine groups of some proteins and nitrites [63]. The already produced nitrosamines that are exogenous in nature, are traced in preserved foods, meat products and also in foods produced for drying by additives like malt in formation of whiskey and beer, salty and pickled preserved food components [64].

Salted Fish: Preparation Methods

In Japan, China; etc Asian countries, almost 20 types of fish are taken for the production of fish pickles or various salted fishes. It is considered as a tradition or an Asian delicacy [65].

These salted or pickled fishes are produced by various methods that are listed below:

- Brining
  The fish is kept inside a crude salt water solution till the tissues of the fishes undergoes absorption of the amount of salt which is desired [66].
- Dry Salting
  The various types of fishes undergo intense mixing along with dry salt until a brine is produced that is drained away intentionally [67]. Brine is a solution produced because of the dissolved salt present in the water which is there in the fish.
- Pickle curing
  A mixture of the salt and the fish is taken for storage and this is how brine is produced [68]. The entire procedure is done in open air. The mixture is also kept open under the air throughout the entire process but the time of keeping the mixture out open, depends on the type of preparation of the fish, the type of fish used until the needed level of pungent smell and required odour is received [69].

Carcinogenic Food Colours

Allura-Red

Allura Red chemically named as disodium 2-hydroxy-1-(2-methoxy-5-methyl-4 sulphonatophenylazo) naphthalene-6-sulphonate was first introduced in the USA as a food colourant [70, 71]. It is non-genotoxic in many gene mutation tests involving prokaryotic and eukaryotic cells with or without activation. However, Allura red sometimes show direct genotoxic effect when different concentrations of the dye ranging from 9.76 to 5000 μg/mL was incubated with a culture of *Saccharomyces cerevisiae* at 37°C

[72]. Comet assay revealed dose-related DNA damage starting at concentration of 1250 μg/ml, though no positive correlation could be established with [73].

In Vivo Assays

An experiment was conducted to show the non-teratogenicity of Allura red [74]. Groups of 11-day old pregnant rats were fed up to 2000 mg/kg b.w. of single oral doses of Allura red [75]. The comet assay (single-cell gel electrophoresis) was used for measuring deoxyribonucleic acid (DNA) strand damage in the embryo at 3, 6 and 24 h sampling times. Finally, it was observed that there was a colon-specific DNA damage in mice at doses of 10 mg/kg b.w., three-hour post-administration however no significant damage was detected in rats at any of the tested dye doses or exposure [76]. By using a flow-cytometric based micronucleus assay in vivo Abramsson and Ilbäck (2013) reported that intra-peritoneal administration of Allura red is not responsible for causing depression in cell proliferation or an increase in the frequency of micronuclei in polychromate erythrocytic cells [77]. The authors tested the dyes up to doses of 2000 mg/kg b.w., checked the use of the intra-peritoneal route on the poor uptake of Allura red from the intestine and finally concluded that this together with the possibility of microbial formation of substances with local cellular action is responsible for the toxicity seen in the report [78].

Sunset Yellow FCF

Disodium 6-hydroxy-5-[(4-sulfophenyl) azo]-2-naphthalenesulfonate is the Chemical name. There were several reports showing the non-mutagenicity of sunset yellow on *E. coli* and four tester strains of *S. typhimurium* TA 1538, 1535, 100 and 98 strains with or without metabolic activation [79, 80]. It was proved by Haveland-Smith and Combes in 1980 that twenty-five dyes have the ability to induce mutations in a tryptophan-requiring *E. coli* strain (sensitive to base substitutions) and a histidine auxotroph of *S. typhimurium* strain TA1538 (specific for frameshifts) (Gerhäuser et al., 2003). Finally, Sunset yellow was concluded to be a non-mutagenic with or without metabolic activation (Peers and Linsell et al., 1973) [81]. An experiment was conducted in vivo where a comet assay was succeeded by a single oral dosing of the dye up to 2000 mg/kg and finally it was concluded that Sunset Yellow FCF did not result in DNA damage in any of the eight mouse organs-glandular stomach, colon, liver, kidney, urinary bladder, lung, brain, bone marrow [82]. HPLC was used in monitoring the metabolites in faeces and were reported to be non-genotoxic as well [83].

Chronic/Long-Term Carcinogenicity Studies

A report had been established showing the administration of sunset yellow in doses of 0 to 2% in mice for periods of 52 to 80 weeks with a conclusion that there is significant difference in the incidence of tumours when compared with other appropriate control groups JECFA [84, 85]. Similar long-term studies in rats, hamster and dogs have not detected any carcinogenic effects associated with the dye when administered up to 5% dose [86].

Brilliant Blue FCF

It is best known with the chemical name: Ethyl - [4 - [[4 - [ethyl - [(3 - sulfophenyl) methyl] amino] phenyl] - (2 - sulfophenyl) methylidene] - 1-cyclohexa - 2, 5 - dienylidene]-[(3-sulfophenyl) methyl] azanium [87].

Gene Mutation Assay

The non-mutagenicity of brilliant blue in various *Salmonella* strains with or without metabolic activation has been demonstrated in many studies [88].

In Vivo Assays

In a study it was found that Brilliant Blue is not responsible for increasing the frequency of micronuclei in the bone marrow of groups of mice receiving intra-peritoneal doses of 0, 500, 1000 or 2000 mg/kg b.w [89]. In another study brilliant blue was reportedly genotoxic in the chromosome aberration assay of over 190 additives using Chinese hamster fibroblast cell line because of the inconsistent osmolality of cell culture medium. At doses up to 2000 mg/kg b.w. the colourant gave negative results in a genotoxicity assessment of 39 food additives by comet assay on eight mouse organs [90, 91]. Dye doses of 0, 0.1, 1.0 or 2.0% were fed to groups of Charles River CD rats for a period of 116 weeks for the males and 111 weeks for the females in a lifetime/carcinogenicity study involving F0 and F1 [92]. A no-observed-adverse-effect-level (NOAEL) of 2 and 1% was established for the male and female groups respectively. In the same study no adverse effects were observed with CD mice when fed up to 5% dye in addition to that NOAEL was established at 5% for both genders [93]. It was observed in a study that dietary concentrations of brilliant Blue, 0.03, 0.3 or 3% when fed to groups of 30 rats for 75 weeks did not produce any dose-related adverse effects on the growth or mortality [94]. Similar results were also seen with the sub-cutaneous injection [95].

Different Preservatives Used

There are several other agents found in food and other consumer goods along with colourants [96]. Preservatives in such food are mainly used to prevent or retard spoilage caused by chemical changes, e.g., oxidation or the growth of mold [97, 98]. Safety concerns are also equally important in the use of these preservatives. The following section will provide an analysis of some commonly used, approved preservatives along with some tests that have been accomplished to determine their toxicity or otherwise [99, 100, 101].

Tartrazine

It is known by the Commercial name- 3-carboxy-5-hydroxy-1-(4'sulpho-phenyl)-4-(4' sulphophenylazo) pyrazole trisodium salt.

Cytogenetic Evaluation of Tartrazine

In a recent study it was found that the interaction of tartrazine and endogenous material like bovine haemoglobin spontaneously involve Van der Waal's forces and hydrogen bonds between the oxygen atoms at position 31 and 15 in the dye [102, 103]. Likewise, tartrazine showed extensive DNA binding, cytostatic potential and reduced mitotic index in another similar study [104].

Gene Mutation Assays

The non-mutagenicity of tartrazine using *Salmonella typhimurium* and *Escherichia coli* was specified by several reports [105]. Even the addition of rat liver microsome preparation S9 have also not shown the colourant as genotoxic [106]. The inability of the dye (at 10 to 1000 µg/plate concentration) to induce point mutation in Tester strain T97a and base pair mutation in T100 in a *Salmonella* [107]. However, the strain T98 showed a significant result but this was not dose related [108, 109].

In Vivo Assays

The results of comet assay on eight mouse organs were recorded following oral administration of tartrazine up to doses of 2000 mg/kg b.w. DNA damage was reported at dose levels of 10 mg/kg b.w in the colon and the glandular stomach at doses higher than 10 mg/kg b.w. only after 3 h post-administration [110, 111]. It thus demonstrated the no mutagenicity of tartrazine when administered as oral gavage up to doses of 2000 mg/kg b.w. The dye did not produce an increase in the number of micronucleate colonic cells at any of the tested doses when compared with water and olive oil control groups [112]. However, it produced an increment in mitotic index even at all tested doses although this effect was not dose related [113]. It was thereafter concluded that the observed extensive DNA damage at the same dose level was due to the acute cytotoxicity of the dye [114]. It could also be due to insufficient DNA repair at the 3h sampling time [115].

Chronic Toxicity

Long term carcinogenicity tests a 90-day oral sub-chronic toxicity involving daily dosing of Wistar rats has been conducted in order to evaluate the effects of tartrazine on chronic administration [116]. A diet containing 5, 7.5 and 10 mg/kg b.w. was fed to the rats divided into groups of six. Then a comparison of the haematological, clinical chemistry and histopathology parameters of the treated animals were conducted with a control group that received only distilled water [117]. The results showed significant dose-related increase in blood chemistry parameters like glucose, creatinine, blood urea nitrogen, total cholesterol, total serum protein, triglycerides and aspartate aminotransferase when compared with the control group [118, 119]. When tartrazine was administered in doses of 0, 1 and 2% in the drinking water of groups of 50 male and female rats for 2 years a lack of carcinogenicity in F334 rats have been reported [120]. No toxic lesions were reported in the treated groups at any of the dye doses. In both the treated and control groups of this strain of rats, all of the tumours observed were characteristically spontaneous [121, 122]. No difference was found in the organ distribution and the histology of the affected organs when treated groups were compared with the control [123]. Similarly, derived from historical controls when the incidences of spontaneous mesothelioma (amongst males) were compared with that of endometrial stromal polyps (in females) it was concluded that the increase in the incidences of these tumours that was observed in the 1% treated group is not related to the dye administration [124].

Reproduction/Developmental-Toxicology

A report has been made carrying the teratogenic potential of dose levels of 0, 60, 100, 200, 400 and 600 mg/kg b.w of tartrazine when administered to pregnant Osborne-Mendel within the first 19 days of gestation [125]. Tartrazine did not produce any significant toxic effect in the treatment groups when compared with the control groups when presented either as oral gavages or in drinking water [126]. Clinical findings, rate/success of implantation, foetal size and development, foetal bone and visceral organ development resulted in the primary outcomes which were used to assess teratogenicity [127]. However, toxicity of tartrazine on the reproductive organs of male Swiss albino mice which were administered doses of 0, 0.1, 1.0 and 2.5% in drinking water for 13 weeks were reported in another study [128]. The results showed that although there was no significant difference in the weights of the seminal vesicle and epididymis of the treated animals, a reduced sperm count and increased incidence of sperm abnormalities were observed in the treatment group that received 2.5% dose. They also reported a reduction in reproductive performance as estimated by mating index in the groups that received 1 and 2.5% doses [129]. Hyperactivity, anxiety and depression effects Tartrazine was also reported for causing an increased activity, anxiety and anti-social behaviour among male Wistar rats [130, 131]. An experiment was conducted where tartrazine doses of 0, 1 and 2.5% mixed in drinking water was given to the rats in three groups of 15. An increase in anxiety level was measured by the different animal models of open field, elevated plus-maze and the dark-light transition tests was observed in the treated animals showed [132] hyperactivity, as measured with the open maze model, depression and a decreased frequency of social contacts in the treated groups was also reported [133, 134, 135].

Conclusion

The above review establishes a relation between development of cancer and few food items or constituents of food items. Even, there are some food items which have many notable dietary inputs and even they can result in the development of cancer [136]. According to many recent studies, dietary rules should be followed strictly for maintaining a healthy life [137]. But also, many researches and studies have been conducted to ensure that the risk in taking these dietary food items is reduced to a great extent and also to ensure that, the carcinogenic elements present in these food items are removed with great care so that no trace of carcinogenicity can be found in these food items [138]. Only by this way, we can hope for a healthy and a better life ahead of us. General awareness and education from the school level about these carcinogens and carcinogenic food items should be provided to make people more and more aware about carcinogenic food [139]. Experimental results of these carcinogenic components like acrylamide on humans are very less, but more and more experiments are conducted on animals like rodents and mice and observe the carcinogenic changes in them caused by these carcinogenic compounds [140]. For an instance, acrylamide is produced in the production of one of the most popular and most consumed fast food for example potato chips and French fries [141]. But over the time period, many scientific progresses have been made to avoid and remove these carcinogenic components from these food items and then let these products be taken for industrial production [142]. Taken together, a little precaution can definitely lead to a better future without any carcinogenic compounds [143, 144].

Conflicts of Interest

This is an original research article and the authors report no conflicts of interest. The authors alone are responsible for the content and writing of the paper.

References

[1] Abnet, CC, Carcinogenic food contaminants. *Cancer Investigation,* 25 (2007) 189–196.

[2] Ghosh, S, Sarkar, T, Das, A, Chakraborty, R, Natural colorants from plant pigments and their encapsulation: An emerging window for the food industry. *LWT,* 153 (2022), 112527.

[3] Biswas, SK, McClure, D, Jimenez, LA, Megson, IL, Rahman, I, Curcumin induces glutathione biosynthesis and inhibits NF-kappaB activation and interleukin-8

release in alveolar epithelial cells: mechanism of free radical scavenging activity. *Antioxidant Redox Signal,* 7(2005) 32–41.

[4] Davidson, W, Ash, S, Capra, S, Bauer, J, Cancer Cachexia Study Group: weight stabilisation is associated with improved survival duration and quality of life in unresectable pancreatic cancer. *Clinical Nutrition,* 23 (2004) 239–247.

[5] Ghosh, S, Sarkar, T, Das, A, et al. Micro and Nanoencapsulation of Natural Colors: a Holistic View. *Appl Biochem Biotechnol* 193, (2021) 3787–3811.

[6] Fares, NH, El-Sayed, AK, Fine structural changes in the ileum of mice fed on delta-endotoxin-treated potatoes and transgenic potatoes. *Natural Toxins,* 6 (1998) 216–331.

[7] Gerhäuser, C, et al. Mechanism-based in vitro screening of potential cancer chemopreventive agents. *Mutation Research* 523 (2003) 163–172.

[8] Goldenberg, D, Maté: a risk factor for oral and oropharyngeal cancer. *Oral Oncology,* 38 (2002) 646–649.

[9] Hayes, RB, Van Nieuwenhuize, JP, Raatgever, JW, Ten Kate, FJ, Aflatoxin exposures in the industrial setting: an epidemiological study of mortality. *Food and Chemical Toxicology,* 22 (1984) 39–43.

[10] Inoue-Choi, M, Sinha, R, Gierach, GL, Ward, MH, Red and processed meat, nitrite, and heme iron intakes and post-menopausal breast cancer risk in the NIH-AARP Diet and Health Study. *International Journal of Cancer,* 138 (2016) 1609–1618.

[11] Kanaya, N, Adams, L, Takasaki, A, Chen, S, Whole blueberry powder inhibits metastasis of triple negative breast cancer in a xenograft mouse model through modulation of inflammatory cytokines. *Nutrition and Cancer,* 66 (2013) 242–248.

[12] Kaplan, SE, Ozdemir, A, Savas, M, Cancer risk factors and prevention in Turkey. *International Journal of Health Services Research and Policy,* 3 (2018) 143–150.

[13] Ludovici, V, Barthelmes, J, Nagele, MP, Flammer, AJ, Sudano, I, Polyphenols: anti-platelet nutraceutical? *Current Pharmaceutical Design,* 24, (2018) 146–157.

[14] McKeague, M, Aptamers for DNA damage and repair. *International Journal of Molecular Sciences,* 18 (2017) 2212.

[15] Michels, KB, Mohllajee, AP, Roset-Bahmanyar, E, Beehler, GP, Moysich, KB, Diet and breast cancer. *Cancer* 109 (S12) (2007) 2712–2749.

[16] Miyata, Y, Matsuo, T, Sagara, Y, Ohba, K, Ohyama, K, Sakai, H, A mini-review of reactive oxygen species in urological cancer: correlation with NADPH oxidases, angiogenesis, and apoptosis. *International Journal of Molecular Science,* 18 (2017) 2214.

[17] Miyata, Y, Shida, Y, Hakariya, T, and Sakai, H, Anti-cancer effects of green tea polyphenols against prostate cancer. *Molecules,* 24 (2019) 193.

[18] Muir, DCG, Ford, CA, Grift, NP, Stewart, REA, Bidleman, TF, Organochlorine contaminants in narwhal (Monodon monoceros) from the Canadian Arctic. *Environmental Pollution,* 75 (1992) 307–316.

[19] Oliver, CP, Gloria, MBA, Barbour, JF, Scanlan, RA, Nitrate, nitrite and volatile nitrosamines in whey-containing food products. *Journal of Agricultural and Food Chemistry,* 43 (1995) 967–969.

[20] Peers, FG, Linsell, CA, Dietary aflatoxins and liver cancer–a population-based study in Kenya. *British Journal of Cancer,* 27 (1973) 473–484.

[21] Pitot, CP, Loeb, DL, Fundamentals of Oncology. Marcel Dekker, New York, NY Pöschl, G, Seitz, HK. Alcohol and cancer. *Alcohol and Alcoholism,* 39 (2004) 155–165.

[22] Roswall, N, Weiderpass, E, Alcohol as a risk factor for cancer: existing evidence in a global perspective. *Journal of Preventive Medicine & Public Health,* 48 (2015) 1–9.

[23] Şen, S, Altınkaynak, S, Genetically modified foods and potential health risks. *The SAU Journal of Science,* 18 (2014) 31–38.

[24] Sevim, S, Kızıl, M, Besin karsinojenlerinin detoksifikasyonunda alternatif yöntem: probiyotikler. *Food and Health,* 5(2019) 139–148.

[25] Frankos VH and Rodricks JV. Food additives and nutritionsupplements. In: Gad S (ed.) *Regulatory Toxicology,* 2nd ed. Boca Raton, FL: Taylor and Francis, 2002, 133–166.

[26] Kruger CL, Reddy CS, Conze DB, et al. Food safety andfood-borne toxicants. In: Hayes AW and Kruger CL (eds.) *Hayes' Principles and Methods of Toxicology,* 6th ed. Boca Raton, FL: CRC Press, 2014, 621–675.

[27] US FDA. Department of Health and Toxicological principles for the safety assessment of direct food additives and color additives used in food *Redbook II.* Washington, DC: Center for Food Safety and Applied Nutrition, 1993.

[28] US FDA. Guidance to the industry: preparation of premarket submissions for food contact substances: chemistry recommendations. *Guidance Document. U.S. Food and Drug Administration, 2007.*

[29] US FDA. Preparation of premarket notifications for food contact substances: chemistry recommendations. *Guidance Document. Washington, DC: Center for Food Safety and Applied Nutrition, 1999a.*

[30] US FDA. Indirect food additives: paper and paperboard components, (21CFRPart176). *FedReg.*Vol.81, No.1/Monday, 4 January 2016, Rules and Regulations, 2016, 5–6.

[31] USDA. Continuing study of food Intakes by individuals (CSFII) of the diet and health knowledge survey (DHKS), 1994–1996. *United States Environmental Protection Agency*. Beltsville, MD: Beltsville Hum Nutrition Center, 1997.

[32] *Report of the Diet and Nutrition Survey*. Appendix J: Number and pattern of recording days and the effect of weighting. Her Majesty's Stationary Office (HMSO), London, England, 1995, 345–347.

[33] US FDA. Preparation of premarket notifications for food contact substances: toxicology recommendations. *Guidance Document. Washington, DC: Center for Food Safety and Applied Nutrition, 1999b.*

[34] US FDA. Preparation of premarket notifications for food contact substances: toxicology recommendations. *Guidance Document. Washington, DC: Center for Food Safety and Applied Nutrition, 1988.*

[35] Cassidy K, and Elyashiv-Barad S. US FDA's revisedconsumption factor for polystyrene used in food-contact applications. *Food Addit Contam* 24 (2007) 1026–1031.

[36] Kokoski CJ, Henry SH, Lin CS, et al. Method used in safetyevaluation. In: Branen AL, Davidson PM and Salminen S, et al. (eds.) *Food Additives.* NewYork: Marcel Dekker, (1990) 579–616.

[37] Munro IC. Issues to be considered in the safety evaluation offat substitutes. *Food Chem Toxicol* 28 (1990) 751–753.

[38] Dorne JLCM and Renwick AG. The refinement of uncertainty/safety factors in risk assessment by incorporation of data on toxicokinetic variability in humans. *Toxicol Sci* 86 (2005) 20–26.

[39] Kruger CL, Booth N and Hayes AW. Policing ourselves: isGRAS robust enough? *Food Technol* 65 (2011) 18–19.

[40] SCOGS (Select Committee on GRAS Substances). Evaluation of health aspects of GRAS food ingredients: lessons learned and questions unanswered. *Fed Proc* 36 (1981) 2519–2562.

[41] IFT (Institute of Food Technologists Foundation). Makingdecisions about the risk of chemicals in foods with limited scientific information. *Compr Rev Food Sci F* 8 (2009) 269–303.

[42] Kruger CL and Mann SW. Safety evaluation of functionalingredients. *Food Chem Toxicol* 41 (2003)793–805.

[43] Abdel-Rahman A, Anyangwe N, Carlacci L, et al. The safetyand regulation of natural products used as foods and food ingredients. *Toxicol Sci* 123 (2011) 333–348.

[44] Bawa AS and Anilakumar KR. Genetically modified foods: safety, risks and public concerns—a review. *J Food Sci Technol* 50(6) (2013) 1035–1046.

[45] Gruissem W. Genetically modified crops: the truth unveiled. *Agri Food Sec* 4 (2015) 3–10.

[46] Codex Alimentarius Commission. Appendix III, Guideline for the conduct of food safety assessment of foods derived from recombinant-DNA plants and Appendix IV, Annex on the assessment of possible allergenicity. *Alinorm 03/34: JointFAO/WHO Food Standard Programme, Codex Alimentarius Commission, Twenty-Fifth Session, Rome, Italy, 30 June–5 July 2003*, 47–60.

[47] EFSA. Guidance document of the scientific panel on genetically modified organisms for the risk assessment of genetically modified plants and derived food and feed. *EFSA J* 99 (2006a) 1–100.

[48] EFSA. Opinion of the scientific panel on genetically modified organisms on an application (Reference EFSA- GMOUK-2005-14) for the placing on the market of genetically modified potato EH92-527-1 with altered starch composition, for production of starch and food/feed uses, under Regulation (EC) No 1829/2003 from BASF Plant Science. *EFSA J* 324 (2006b) 1–20.

[49] EFSA. Opinion of the scientific panel on genetically modified organisms on a request from the commission related to the notification (Reference CE/ES/00/01) for the placing on the market of herbicide-tolerant genetically modified maize NK603, for import and processing, under Part C of Directive 2001/18/EC from Monsanto (Question No EFSA-Q-2003003). *EFSA J* 10 (2003) 1–13.
[50] EFSA. Opinion of the scientific panel on genetically modified organisms on a request from the commission related to the notification (Reference C/NL/00/10) for the placing on the market of insect-tolerant genetically modified maize 1507, for import and processing, under Part C of Directive 2001/18/EC from pioneer hi-bred international/mycogen seeds. *EFSA J* 124 (2004a) 1–18.
[51] EFSA. Application (Reference EFSA-GMO-NL-2007-37) forthe placing on the market of the insect-resistant genetically modified maize MON89034, for food and feed uses, import and processing under Regulation (EC) No 1829/2003 from Monsanto. *EFSA J* 909 (2008a) 1–30.
[52] EFSA. Opinion of the scientific panel on genetically modified organisms on application (reference EFSA- GMONL-2006-36) for the placing on the market of the glyphosate-tolerant genetically modified soybean MON89788, for food and feed uses, import and processing under Regulation (EC) No 1829/2003 from Monsanto. *EFSA J* 758 (2008b) 1–23.
[53] EFSA. Application (Reference EFSA-GMO-CZ-2005-27) forthe placing on the market of the insect-resistant and herbicide-tolerant genetically modified maize MON88017, for food and feed uses, import and processing under Regulation (EC) No 1829/2003 from Monsanto. *EFSA J* 1075 (2009a) 1–28.
[54] EFSA. Application (ReferenceEFSA-GMO-UK-2005-11) forthe placing on the market of insect-resistant genetically modified maize MIR604 event, for food and feed uses, import and processing under Regulation (EC) No 1829/2003 from Syngenta Seeds S.A.S on behalf of Syngenta Crop Protection AG. *EFSA J* 1193 (2009b) 1–26.
[55] EFSA. Applications (EFSA-GMO-RX-MON810) for renewal of authorization for the continued marketing of (1) existing food and food ingredients produced from genetically modified insect resistant maize MON810; (2) feed consisting of and/or containing maize MON810, including the use of seed for cultivation; and of (3) food and feed additives, and feed materials produced from maize MON810, all under Regulation (EC) No 1829/2003 from Monsanto. *EFSA J* 1149 (2009c) 1–85.
[56] EFSA. Applications (References EFSA-GMO-NL-2005-22, EFSA-GMO-RXNK603) for the placing on the market of the genetically modified glyphosate tolerant maize NK603 for cultivation, food and feed uses, import and processing and for renewal of the authorization of maize NK603 as existing products, both under Regulation (EC) No 1829/2003 from Monsanto. *EFSA J* 1137 (2009d) 1–50.
[57] EFSA. Scientific Opinion on application (EFSA-GMO-NL2005-16) for the placing on the market of insect resistant genetically modified cotton (Gossypiumhirsutum L.) 28124-236x3006-210-23 for food and feed uses, import and processing under Regulation (EC) No 1829/2003 from Dow Agro Sciences. *EFSA J* 8(6) (2010a) 1644.

[58] EFSA. Scientific opinion on applications (EFSA-GMO-RX40-3-2[8-1a/20-1a], EFSA-GMO-RX-40-3-2[8-1b/20-1b]) for renewal of authorization for the continued marketing of (1) food containing, consisting of, or produced from genetically modified soybean 40-3-2; (2) feed containing, consisting of, or produced from soybean 40-3-2; (3) other products containing or consisting of soybean 40-3-2 with the exception of cultivation, all under Regulation (EC) No 1829/2003 from Monsanto. *EFSA J* 8(12) (2010b) 1908.

[59] EFSA. Scientific Opinion on application (EFSA-GMO-BE2010-79) for the placing on the market of insect resistant genetically modified soybean MON 87701 for food and feed uses, import and processing under Regulation (EC) No 1829/ 2003 from Monsanto. *EFSA J* 9(7) (2011a) 2309.

[60] EFSA. Scientific opinion on application (EFSA-GMO-CZ2008-54) for placing on the market of genetically modified insect resistant and herbicide tolerant maize MON 88017 for cultivation under Regulation (EC) No 1829/2003 from Monsanto. *EFSA J* 9(11) (2011b) 2428.

[61] EFSA. Scientific opinion on application (EFSA-GMO-NL2008-52) for the placing on the market of herbicide tolerant genetically modified soybean A5547- 127 for food and feed uses, import and processing under Regulation (EC) No 1829/2003 from Bayer crop science. *EFSA J* 9(5) (2011c) 2147.

[62] EFSA. Scientific opinion on application (EFSA-GMO-UK2007-43) for the placing on the market of herbicide tolerant genetically modified soybean 356043 for food and feed uses, import and processing under regulation (EC) No 1829/2003 from pioneer. *EFSA J* 9(7) (2011d) 2310.

[63] EFSA. Final review of the Se´ral inietal. (2012a) publicationon a 2-year rodent feeding study with glyphosate formulations and GM maize NK603 as published online on 19 September 2012 in Food and Chemical Toxicology. *EFSA J* 10(11) (2012a) 2986.

[64] EFSA. Scientific opinion on an application (EFSA-GMONL-2005-24) for the placing on the market of the herbicide tolerant genetically modified soybean 40-3-2 for cultivation under regulation (EC) No 1829/2003 from Monsanto. *EFSA J* 10(6) (2012b) 2753.

[65] EFSA. Scientific opinion on an application (EFSA-GMONL-2009-70) for the placing on the market of genetically modified drought tolerant maize MON 87460 for food and feed uses, import and processing under regulation (EC) No 1829/2003 from Monsanto. *EFSA J* 10(11) (2012c) 2936.

[66] EFSA. Scientific opinion on application (EFSA-GMO-NL2010-78) for the placing on the market of herbicidetolerant, increased oleic acid genetically modified soybean MON 87705 for food and feed uses, import and processing under regulation (EC) No 1829/2003 from Monsanto. *EFSA J* 10(10) (2012d) 2909.

[67] Goodman RE, Vieths S, Sampson HA, et al. Allergenicityassessment of genetically modified crops—What makes sense? *Nat Biotechnol* 26(1) (2008) 73–81.

[68] Magnuson B, Munro I, Abbot P, et al. Review of the regulation and safety assessment of food substances in various countries and jurisdictions. *Food Addit Contam A* 30 (2013) 1147–1220.

[69] Blaauboer BJ, Boobis AR, Bradford B, et al. Consideringnew methodologies in strategies for safety assessment of foods and food ingredients. *Food Chem Toxicol* 91 (2016) 19–35.
[70] Leist M, Lidbury BA, Yang C, et al. Novel technologies andan overall strategy to allow hazard assessment and risk prediction of chemicals, cosmetics, and drugs with animal-free methods. ALTEX 29(4) (2012) 373–388.
[71] Leist M and Hartung T. Inflammatory findings on speciesextrapolations: humans are definitely no 70-kg mice. *Arch Toxicol* 87(4) (2013) 563–567.
[72] Suter-Dick L, Alves PM, Blaauboer BJ, et al. Stem cellderived systems in toxicology assessment. *Stem Cells Dev* 24(11) (2015) 1284–1296.
[73] Hartung T, Luechtefeld T, Maertens A, et al. Integrated testing strategies for safety assessments. ALTEX 30(1) (2013b) 3–18.
[74] Dollinger M. & Rosenbaum E. *Everyone's Guide to Cancer Therapy: How Cancer is Diagnosed, Treated and Managed Day to Day*. (5th Edition). Kansas City: Andrews McMeel Publishing, 19 (2008) 155-164.
[75] Ecsedy J & Hunter D. The origin of cancer. Adami, H.-O., Hunter, D., &Trichopoulos, D. *Textbook of Cancer Epidemiology*. (2nd Edition). Oxford: Oxford University Press, 3 (2008) 61-85.
[76] Erson AE & Petty EM. Molecular and genetic events in neoplastic transformation. Schottenfeld, D. &Fraumeni, J. F. Jr. (Eds.). *Cancer Epidemiol. Prev.* (3rd Edition). New York: Oxford University Press, 4 (2006) 47-64.
[77] H & Khayat D et al. (Eds.). *UICC Manual of Clinical Oncology*. (8th Edition). New Jersey: John Wiley & Sons Inc., (2004) 1-18.
[78] Merkle CJ. Biology of cancer. Yarbro CH, Wujcki D, Holmes Gobel B (Eds.). *Cancer Nursing: Principles and Practice.* (7th Edition). Sudbury, MA: Jones and Bartlett, 1 (2011) 3-22.
[79] Mills GB, & Trahan Rieger P. Genetic predisposition to cancer. Pollock RE, Doroshow J H & Khayat D et al. (Eds.). *UICC Manual of Clinical Oncology*. (8th Edition). New Jersey: John Wiley & Sons, Inc. 4 (2004) 63-89.
[80] Ruccione K. Biologic Basis of Cancer in Children and Adolescents. Baggott CR, Kelly K P, Fochtman D et al. *Nursing Care of Children and Adolescents with Cancer.* (3rd Edition). Philadelphia, PA: W. B. Saunders Company, 2 (2002) 24 – 63.
[81] Ledford H. Disputed definitions. *Nature,* 455 (2008) 1023–8.
[82] Razin A, Cedar H. DNA methylation and gene expression. *Microbiol Rev,* 55(3) (1991) 451-458.
[83] Strahl BD, Allis CD. The language of covalent histone modifications. *Nature,* 403, (2000) 41–45.
[84] Jenuwein T, Allis CD. Translating the histone code. *Science,* 293 (2001) 1074–1080;
[85] Bergmann A, Lane ME. Hidden targets of microRNAs for growth control. *Trends Biochem Sci,* 28 (2003) 461–463.
[86] Mottram, DS, Wedzicha, BL and Dodson, AT, Acrylamide is formed in the Maillard reaction. *Nature,* 419(6906) (2002) 448–449.

[87] Kawamura S. Seventy years of the Maillard reaction. In *The Maillard Reaction in Foods and Nutrition*. Waller, G. R. and Feather, M. S, (Eds.). ACS Symp. Ser, (1983) 3–18.

[88] Gilcrest, L, FDA surveys show eight foods account for bulk of acrylamide in US diet. *Food Chemical News,* 45(3) (2003 b) 26-27.

[89] Mucci L, Dickman P, Steineck G, et al. Dietary acrylamide and cancer of the large bowel, kidney, and bladder: absence of an association in a population-based study in Sweden *Br J Cancer,* 88 (2003) 84-89.

[90] Mucci L, Sandin S, Balter K, et al. Acrylamide intake and breast cancer risk in Swedish women. *JAMA,* 293(11) (2005) 1326-1327.

[91] Coppin JF, Qu W, Waalkes MP. Interplay between cellular methyl metabolism and adaptive efflux during oncogenic transformation from chronic arsenic exposure in human cells. *J Biol Chem.* 283 (2008) 19342–19350.

[92] Kitchin KT, Ahmad S. Oxidative stress as a possible mode of action for arsenic carcinogenesis. *Toxicol Lett.* 137 (2003) 3–13.

[93] Kojima C, Ramirez DC, Tokar EJ, Himeno S, Drobná Z, Stýblo M, Mason RP, Waalkes MP. Requirement of arsenic biomethylation for oxidative DNA damage. *J Natl Cancer Inst*. 101 (2009) 1670–1681.

[94] Valinluck V, Sowers LC. Endogenous cytosine damage products alter the site selectivity of human DNA maintenance methyltransferase DNMT1. *Cancer Res.* 67 (2007) 946–950.

[95] Lee DH, Jacobs DR, Jr, Porta M. Hypothesis: a unifying mechanism for nutrition and chemicals as lifelong modulators of DNA hypomethylation. *Environ Health Perspect.* 117 (2009) 1799–1802.

[96] Mass MJ, Wang L. Arsenic alters cytosine methylation patterns of the promoter of the tumor suppressor gene p53 in human lung cells: a model for a mechanism of carcinogenesis. *Mutat Res.* 386 (1997) 263–277.

[97] Huang YC, Huang WC, Chen WT, Yu HS, Chai CY. Sodium arsenite-induced DAPK hypermethylation and autophagy via ERK1/2 phosphorylation in human uroepithelial cells. *Chem Biol Interact.* 181 (2009) 254–262.

[98] Zhou X, Sun H, Ellen TP, Chen H, Costa M. Arsenite alters global histone H3 methylation. *Carcinogenesis.* 29 (2008) 1831–1836.

[99] Beezhold K, Liu J, Kan H, Meighan T, Castranova V, Shi X, Chen F. miR-190-mediated downregulationof PHLP contributes to arsenic-induced Akt activation and carcinogenesis. *Toxicol Sci.* 2011 in press.

[100] Salnikow K, Zhitkovich A. Genetic and epigenetic mechanisms in metals carcinogenesis and cocarcinogenesis: nickel, arsenic, and chromium. *Chem Res Toxicol.* 21 (2008) 28–44.

[101] Martinez-Zamudio R, Ha HC. Environmental epigenetics in metal exposure. *Epigenetics.* (2011) 820–827.

[102] IARC. Monographs on the Evaluations of Carcinogenic Risks to Humans. 1,3-Butadiene, ethylene oxide and vinyl halides (vinyl fluoride, vinyl chloride, and vinyl bromide) *World Health Organization International Agency for Research on Cancer*, Vol. 97. IARC; Lyon, France: 2008.

[103] Melnick RL, Sills RC. Comparative carcinogenicity of 1,3-butadiene, isoprene and chloroprene in rats and mice. *Chem Biol Interact* 136. 136 (2001) 27–42.

[104] Walker VE, Walker DM, Meng Q, McDonald JD, Scott BR, Selikop SK, Claffey DJ, Upton PB, Powley MW, Swenberg JA, Henderson RF, Health Review Committee Genotoxicity of 1,3-butadiene and its epoxy intermediate. *Res Rep Health Eff Inst.* 144 (2009) 3–79.

[105] Koturbash I, Scherhag A, Sorrentino J, Sexton K, Bodnar W, Tryndyak V, Latendresse JR, Swenberg JA, Beland FA, Pogribny IP, Rusyn I. Epigenetic alterations in liver of C57BL/6J mice after short-term inhalational exposure to 1,3-butadiene. *Environ Health Perspect.* 119 (2011) 635–640.

[106] Schotta G, Lachner M, Sarma K, Ebert A, Sengupta R, Reuter G, Reinberg D, Jenuwein T. A silencing pathway to induce H3-K9 and H4-K20 trimethylation at constitutive heterochromatin. *Genes Dev.* 18 (2004) 1251–1262.

[107] Martens JH, O'Sullivan RJ, Braunschweig U, Opravil S, Radolf M, Steinlein P, Jenuwein T. The profile of repeat-associated histone lysine methylation states in the mouse epigenome. EMBO J. 24 (2005) 800–812.

[108] Jenuwein Y. The epigenetic magic of histone lysine methylation. *FEBS J.* 273 (2006) 3121–3135.

[109] Koturbash I, Scherhag A, Sorrentino J, Sexton K, Bodnar W, Swenberg JA, Beland FA, Pardo-Manuel Devillena F, Rusyn I, Pogribny IP. Epigenetic mechanisms of mouse interstrain variability in genotoxicity of the environmental toxicant 1,3-butadiene. *Toxicol Sci.* 122 (2011) 448–456.

[110] Rusyn I, Gatti DM, Wilshire T, Kleeberger SR, Threadgill DW. Toxicogenomics: population-based testing of drug and chemical safety in mouse models. *Pharmacogenomics.* 11 (2010) 1127–1136.

[111] IARC Monographs on the Evaluations of Carcinogenic Risks to Humans . A review of human carcinogens. *Part A: Pharmaceuticals.* Vol. 100. IARC; Lyon, France: 2011.

[112] Li S, Washburn KA, Moore R, Uno T, Teng C, Newbold RR, McLachlan JA, Negishi M. Developmental exposure to diethylstilbestrol elicits demethylation of estrogen-responsive lactoferrin gene in mouse uterus. *Cancer Res.* 57 (1997) 4356–4359.

[113] Li S, Hansman R, Newbold R, Davis B, McLachlan JA, Barrett JC. Neonatal diethylstilbestrol exposure induces persistent elevation of c-fos expression and hypomethylation in its exon-4 in mouse uterus. *Mol Carcinog.* 38 (2003) 78–84.

[114] Tang WY, Newbold R, Mardilovich K, Jefferson W, Cheng RY, Medvedovic M, Ho SM. Persistent hypomethylation in the promoter of nucleosomal binding protein 1 (Nsbp1) with overexpression of Nsbp1 in mouse uteri neonatally exposed to diethylstilbestrol or genistein. *Endocrinology.* 149 (2008) 5922–5931.

[115] Sato K, Fukata H, Kogo Y, Ohgane J, Shiota K, Mori C. Neonatal exposure to diethylstilbestrol alters expression of DNA methyltransferases and methylation of genomic DNA in the mouse uterus. *Endocr J.* 56 (2009)131–139.

[116] Hsu PY, Detherage DE, Rodriguez BA, Liyanarachchi S, Weng YI, Zuo T, Liu J, Cheng AS, Huang TH. Xenoestrogen-induced epigenetic repression of microRNA-9-3 in breast epithelial cells. *Cancer Res.* 69 (2009) 5936–5945.

[117] Lehmann U, Hasemeier B, Christgen M, Müller M, Römermann D, Länger F, Kreipe H. Epigenetic inactivation of microRNA gene has-mir-9-1 in human breast cancer. *J Pathol.* 214 (2008) 17–24.

[118] Jordan VC. Tamoxifen (ICI146,474) as a target therapy to treat and prevent breast cancer. *Br J Pharmacol.* 147 (2006) S269–S276.

[119] Shang Y. Molecular mechanisms of oestrogen and SERMs in endometrial carcinogenesis. *Nat Rev Cancer.* 6 (2006) 360–368.

[120] Wu H, Chen Y, Liang J, Shi B, Wu G, Zhang Y, Wang D, Li R, Yi X, Zhang H, Sun L, Shang Y. Hypomethylation-linked activation of PAX2 mediates tamoxifen-stimulated endometrial carcinogenesis. *Nature.* 438 (2005) 981–987.

[121] Wogan GN. Review of the toxicology of tamoxifen. *Semin Oncol.* 24 (1997) S87–S97.

[122] Phillips DH. Understanding the genotoxicity of tamoxifen? *Carcinogenesis.* 22 (2001) 839–849.

[123] Gamboa da Costa G, McDaniel-Hamilton LP, Heflich RH, Margues MM, Beland FA. DNA adduct formation and mutant induction in Sprague-Dawley rats treated with tamoxifen and its derivatives. *Carcinogenesis.* 22 (2001) 130701315.

[124] Tryndyak VP, Muskhelishvili L, Kovalchuk O, Rodriguez-Juarez R, Churchwell MI, Ross SA, Beland FA, Pogribny IP. Effect of long-term tamoxifen exposure on genotoxic and epigenetic changes in rat liver: implications for tamoxifen-induced hepatocarcinogenesis. *Carcinogenesis.* 27 (2006) 1713–1720.

[125] Tryndyak VP, Kovalchuk O, Muskhelishvili L, Montgomery B, Rodriguez-Juarez R, Melnyk S, Ross SA, Beland FA, Pogribny IP. Epigenetic reprogramming of liver cells in tamoxifen-induced rat hepatocarcinogenesis. *Mol Carcinog.* 46 (2007) 187–197.

[126] Pogribny IP, Tryndyak VP, Boyko A, Rodriguez-Juarez R, Beland FA, Kovalchuk O. Induction of microRNAome deregulation in rat liver by long-term tamoxifen exposure. *Mutat Res.* 619 (2007) 30–37.

[127] Bachman AN, Phillips JM, Goodman JI. Phenobarbital induces progressive patterns of GC-rich and gene-specific altered DNA methylation in the liver of tumor-prone B6C3F1 mice. *Toxicol Sci.* 91 (2006) 393–405.

[128] Phillips JM, Goodman JI. Multiple genes exhibit Phenobarbital-induced constitutive active/androstane receptor-mediated DNA methylation changes during liver tumorigenesis and in liver tumors. *Toxicol Sci.* 108 (2009) 273–289.

[129] Lempiäinen H, Müller A, Brasa S, Teo SS, Roloff TC, Morawiec L, Zamurovic N, Vicart A, Funhoff E, Couttet P, Schübeler D, Grenet O, Marlowe J, Moggs J, Terranova R. Phenobarbital mediates an epigenetic switch at the constitutive androstane receptor (CAR) target gene Cyp2b10 in the liver of B6C3F1 mice. *PLoS One.* 6 (2011) e18216.

[130] Bucher JR, Shackelford CC, Haseman JK, Johnson JD, Kurtz PJ, Persing RL. Carcinogenicity studies of oxazepam in mice. *Fundam Appl Toxicol.* 23 (1994) 280–297.

[131] Iida M, Anna CH, Hartis J, Bruno M, Wetmore B, Dubin JR, Sieber S, Bennett L, Cunningham ML, Paules RS, Tomer KB, Houle CD, Merrick AB, Sills RC, Devereux TR. Changes in global gene and protein expression during early mouse liver carcinogenesis induced by non-genotoxic model carcinogens oxazepam and Wyeth-14,643. *Carcinogenesis.* 24 (2003) 757–770.

[132] Lahousse SA, Hoenerhoff M, Collins J, Ton TV, Massinde T, Olson D, Rebolosso Y, Koujitani T, Tomer KB, Hong HH, Bucher J, Sills RC. Gene expression and mutation assessment provide clues of genetic and epigenetic mechanisms in liver tumors of oxazepam-exposed mice. *Vet Pathol.* 48 (2011) 678–699.

[133] Ferguson LR, Philpott M. Nutrition and mutagenesis. *Annu Rev Nutr*. 28 (2008) 313–329.

[134] IARC Monographs on the Evaluations of Carcinogenic Risks to Humans. Some traditional herbal medicines, some mycotoxins, naphthalene and styrene. Vol. 82. IARC; Lyon, France: 2002.

[135] Wild CP, Gong YY. Mycotoxins and human disease: a largely ignored global health issue. *Carcinogenesis.* 31 (2010) 71–82.

[136] Knasmüller S, Bresgen N, Kassie F, Mersch-Sundermann V, Gelderblom W, Zöhrer E, Eckl PM. Genotoxic effects of three Fusarium mycotoxins, fumonisin B1, moniliformin and vomitoxin in bacteria and primary cultures of rat hepatocytes. *Mutat Res*. 391 (1997) 39–48.

[137] Wang JS, Groopman JD. DNA damage by mycotoxins. *Mutat Res.* 424 (1999)167–181.

[138] Zhang YJ, Ahsan H, Chen Y, Lunn RM, Wang LY, Chen SY, Lee PH, Chen CJ, Santella RM. High frequency of promoter hypermethylation of RASSF1A and p16 and its relationship to aflatoxin B1-DNA adduct levels in human hepatocellular carcinoma. *Mol Carcinog.* 35 (2002) 85–92.

[139] Zhang YJ, Chen Y, Ahsan H, Lunn RM, Chen SY, Santella RM. Inactivation of the DNA repair gene O6-methylguanine-DNA methyltransferase by promoter hypermethylation and its relationship to aflatoxin B1-DNA adducts and the p53 mutations in hepatocellular carcinoma. *Int J Cancer.* 103 (2003) 440–444.

[140] Su H, Zhao J, Xiong Y, Xu T, Zhou F, Yuan Y, Zhang Y, Zhuang SM. Large-scale analysis of the genetic and epigenetic alterations in hepatocellular carcinoma from Southeast China. Mutat Res. 641 (2008) 27–35.

[141] Zhao W, Liu H, Liu W, Wu Y, Chen W, Jiang B, Zhou Y, Xue R, Luo C, Wang L, Jiang JD, Liu J. Abnormal activation of the synuclein-gamma gene in hepatocellular carcinomas by epigenetic alteration. Int J Oncology. 28 (2006) 1081–1088.

[142] Kouadio JH, Dano SD, Moukha S, Mobio TA, Creppy EE. Effects of combinations of Fusarium mycotoxins on the inhibition of macromolecular synthesis, malondialdehyde levels, DNA methylation and fragmentation, and viability in Caco-2 cells. *Toxicon.* 49 (2007) 306–317.

[143] Marin-Kuan M, Cavin C, Delatour T, Schilter B. Ochratoxin A carcinogenicity involves a complex network of epigenetic mechanisms. *Toxicon.* 52 (2008) 195–202.

Chapter 5

Food Adulteration: Impacts, Challenges and Interventions

Sonali Paul, Keerti Keswani, Pallabi Nath, Sampreeti Chatterjee, Sanmitro Bhattacherjee and Susmita Mukherjee*

Department of Biotechnology,
University of Engineering and Management,
University Area, New Town, Kolkata, West Bengal, India

Abstract

Food is an essential part of every living organism. It is critical to the development and maintenance of a living organism. However, food products have always been a prey to adulteration practices. Adulteration has been a problem since the dawn of time. To save themselves from the everlasting economic crisis, shopkeepers and dealers often get involved in inappropriate practices. Adulteration is done to increase the output of food products by compromising on the quality. It is often done by the sellers for the sole purpose of increasing profits. Adulterants have fatal effects on health and cause severe genetic mutations, which can be passed down generation after generation. It may seem difficult to detect adulterants with the naked eye, but there are several simple tests that can be performed to determine the purity of the food product purchased. There are different organizations which are trying their level best to keep a check on these malpractices and ensure the security of consumers all over the world.

---

*Corresponding Author's Email: susmita.mukherjee@uem.edu.in.

In: Proceedings of BIOSPECTRUM
Editors: Sanket J. Joshi, Susmita Mukherjee and Moupriya Nag
ISBN: 978-1-68507-985-7

Keywords: food adulteration, foodstuff, adulteration-laws, impact, purity-testing.

Introduction

Adulteration is a severe problem in India that puts many people's health at danger. Adulteration exists in every product. It starts with our everyday shopping and progresses to our life-saving medications. Even infants' milk products have not been spared by the sinners. Food is a vital sustaining factor in man's daily demands. Food preparation is as varied and rich as a man's taste, ranging from the most basic dish to the most sophisticated cuisine. Adulterants have been added to a food as a result of the temptation of money and general apathy for humans, ranging from simple stones in rice to the more damaging brick and boric powder [1].

Adulteration occurs when extraneous substances or adulterants are added to food items or goods, or key nutrients are reduced partially or completely for financial benefit, or owing to negligence and a lack of suitable hygienic conditions during processing, storing, transportation and marketing. As a result, the consumer is either defrauded or is made a frequent victim of ailments. It is critical for consumers to understand prevalent adulterants and their effects on health, as the growing number of foodstuffs imported allows manufacturers to deceive and defraud consumers [2].

What Are Adulterants?

Adulterants can be defined as the purposeful or unintentional addition of a non-food item to increase the quantity of food in raw or prepared form, for preservation purposes, or to improve appearance. Any dangerous or detrimental chemicals that may make the food harmful to health are also listed as adulterants. Adding water to milk or animal carcasses to meat products other than the animal intended for consumption are examples, as are adding stones, gravel sand, insects, or animal hairs to grain goods. Another definition of food adulteration is the purposeful degradation of the quality of food provided for sale [3].

Adulterants are chemical substances which are not meant to be present in our food or beverage. It may be intentionally added to more expensive items to increase the quantity and cut down on manufacturing costs. It may also be done with some malicious intent. The usage of adulterating agents has been a rather common practice in the society since ancient days. These were performed freely due to lack of stringent governing laws and monitoring authorities [4].

"Adulterant" means any material which is or could be employed for making the food unsafe or sub-standard or mis-branded or containing extraneous matter. Food is considered adulterated if a substance that degrades its quality or make it dangerous is added [5].

A food ingredient that has been removed entirely or partially, lowering the product's quality, is known as an adulterated food ingredient. It becomes presentable when contaminated, or it changes colour to improve its appearance. Anything that degrades food quality is either added to it or removed from it. Examples of food adulteration: Mixture of legumes with sand particles, pebbles [6].

Why Is Food Adulterated?

Food items are subjected to adulteration due to several reasons. The most predominant reason being a surge in the prices. By adulteration, shopkeepers and food dealers can easily get more products out of their stock and earn more profits. For example: a milk vendor can make two litres of adulterated product from one litre of pure milk by mixing water. Thus, thc profit from adulterated milk is more than from pure milk. Another reason is to showcase products in a moretempting manner. For example: apples are coated with wax to make them shinier and more appealing to the customer. At times adulteration may not be done intentionally. It may happen due to unsanitary packing surroundings and methods. Some pollutants like brick dust, tar, saw dust etc. may get mixed with the product while packing. Adulteration is also practiced widely during times of economic crisis and supply shortage. Lack of proper preventive measures and stringent laws also play an important role in helping such malpractices to continue [7].

As international population is developing at an alarming rate, meals are regularly adulterated to satisfy the desires of this developing population and to feed the massive scale population.

According to a survey that was taken, adulteration was identified in milk containing 70% water, turmeric powder containing 43% chalk powder, crimson chilli powder containing 100% synthetic colour etc.

The main motivation for adulteration is to increase their monetary flow by increasing the volume of it [9]. Even if some greedy producers, processors, and retailers started adulteration to increase their profit margins, the most important cause of adulteration is dishonesty and a lack of accidental assessment of quality on questionable products [10].

Methods of Adulteration

- Mixing: clay mixing, stone chips, pebbles, sand, marble chips, etc.
- Substitution: Substances that are less expensive and less effective are being partially or completely replaced with better ones.
- Concealing quality: Attempting to conceal the food standard. For example, adding captions of high-quality food to low-quality food for sale [11].
- Decomposed food: It's mostly found in fruits and vegetables. The bad ones are mixed up with the decaying ones.
- Misbranding/False labels: Duplicate foods, changes in manufacturing, and expiration dates are all included.
- Addition of toxicants: non-edible ingredients such as argemone in mustard oil, low-quality preservatives, colouring agents, and so on [12].

Types of Adulterants

Adulteration of foods is done via several methods which can be widely classified into different types: [13]

1. Incidental adulterants - Residues from pesticides, larvae found in foods etc. fall under the category of incidental adulterants.
2. Metallic adulterants- This type of adulterants includes different metals like arsenic, lead, mercury which are mixed with water, pesticides and chemical waste from factory.

3. Intentional adulterants- This type of adulterants include stone, sand, dyes, coal tar etc.
4. Filth and Foreign Adulterants- Filth and extraneous material in food includes foreign matter (for example: glass, metal, plastic, wood, stones, sand, cigarette butts), undesirable parts of raw plant material (for example: stems, pits in pitted olives, pieces of shell in canned oysters), and filth (for example: mould, rot, insect and rodent parts, excreta, decomposition).
5. Economic Adulterant- Adulteration occurs when a food omits or substitutes another substance, in whole or in parts for a valuable constituent (for example: olive oil diluted with tea tree oil); conceals damage or inferiority in any way (for example: food colour is applied on fresh fruit to conceal defects); or any substance has been added to it or packed with it to increase its bulk or whey content (for example: water is added to scallops to make them heavier) [14].
6. Microbial Adulterant- The Presence of pathogens (dangerous microorganisms like bacteria, virus or protozoa) in a food make it adulterated. In general, the presence of virus in ready-to-eat meals contaminates the food. Salmonella can contaminate fresh fruit and vegetables as well as in ready-to-eat meat and poultry product [15].

Table 1. Food items and their respective adulterants

Food item	Adulterant
Mustard seeds	Argemone seeds
Ice cream	Powder detergent
Coffee Powder	Tamarind seeds, chicory powder
Sugar	Chalk powder
Honey	Water/sugar solution

Source: Analysis of Food Adulterants in Selected Food Items Purchased from Local Grocery Stores [17]

Almost every food item ranging from milk to fruits, vegetables to grains, everything has been tempered with in some way. Table 1 gives a brief description on the possible types of adulterants. Some adulterants entering through agricultural stages are not properly cleaned. Stones, leaves, soil, sand and dust are obvious adulterants. The consumer can clean them, making them less harmful. Other intentionally injected adulterants are either undetectable or concealed with colour or texture. They are generally harmful to one's health, and the majority of them result in significant health problems like cancer [16].

Testing of Adulterants

Table 2. Food items and their purity tests

Food Item Type	Purity Tests
	Test 1:
	• Put a drop of milk on a well-polished slanting surface. • In case of pure milk, it either stays or flows down slowly, leaving a white trailbehind. • • Adulterated milk (mixed with water) will flow freely and leave no trace.
Milk and milk products	Test 2:
	• 5 or 10 ml of sample is taken with an equal amount of water. • Mix the contents thoroughly. • If it is adulterated with detergent, it forms dense lather. • When pure milk is agitated, it merely forms a very thin froth layer.
	Test 3:
	• Take 2-3ml of sample and boil with 5ml of water. • Cool it down and add 2-3 drops of tincture of iodine. • Formation of blue colour indicates the presence of starch.
Oils and fats	Test 1:
	• Take coconut oil in a glass. • Refrigerate for 30 minutes. • If it solidifies, then it is pure coconut oil. • If it is adulterated then the other oils present will remain as a separate layer.
	Test 2:
	• 2ml of sample of oil taken. • A little amount of yellow butter is added. • If there is an immediate formation of red colour, presence of Tri-Ortho-Cresyl-Phosphate.
Sugars and confectionary	Test 1:
	• Dip a cotton wick in honey and light it using a flame. • Pure honey burns. • If adulterated, it will produce a crackling sound.
	Test 2:
	• Transparent glass is taken with water in it. • A drop of honey is added into the glass. • Dispersion of honey in water indicates the added sugar.

Food Item Type	Purity Tests
Sugars and confectionary	Test 3:
	• Transparent glass of water is taken. • 10g of sample is dissolved in water. • If chalk is added with sugar, at the bottom adulterant will be seen asprecipitated form.
	Test 4:
	• Some portion of leaves is taken and crushed. • Silver leaves which are pure will be easily crushed into powder whilealuminium ones will break into smaller piece. • Silver leaves which are pure will burn away completely and remaining appearas glistening balls, aluminium leaves are appeared as grey ash.
Salts	Test 1:
	• Burn small amount of asafoetida in a chrome steel spoon. • Pure asafoetida will burn like camphor. • Adulterated asafoetida will now no longer produce brilliant flame likecamphor.
	Test 2:
	• Add a few quantities of black pepper to a tumbler of water. • Pure black pepper settles on the bottom. • In the adulterated black pepper, papaya seeds waft at the floor of water.
	Test 3:
	• Press the berries with the assist of fingers. • Light berries will spoil without difficulty whilst black berries of pepper willnow no longer spoil.
	Test4:
	• Shake little part of the pattern with water and permit to settle. • Pure asafoetida will now no longer go away any cleaning soap stone ordifferent earthy be counted on the bottom. • If asafoetida is adulterated, cleaning soap stone or different earthy be countedwill relax on the bottom.
	Test5:
	• Sprinkle chilli powder at the floor of water taken in a pitcher tumbler. • The synthetic colourants will straight away begin descending in colour streaks.

Source: Detect Adulteration with Rapid Test (DART) [18]

Identifying and extracting adulterants from food items on a large scale is an extremely difficult task. However, over the years, people have been able to discover simple methods to determine the purity level of day-to-day consumables. These tests, as discussed in table 2, can be easily performed at home or any non-industrial level.

Impacts of Adulteration

Adulteration makes the food items we use in our daily life unsafe and unhygienic for consumption. In the past few decades, adulteration of meals has emerged as a severe problem. Intake of adulterated meals cause extreme diseases like cancer, diarrhoea, asthma, ulcers, etc. Adulteration has diverse effects on people from different sectors, from producers/farmers to processors to consumers.

Impacts on Enterprises

Consumer dissatisfaction with their products, product recalls and destruction, complaint expenses, insurance premium hikes, and costs associated with equipment replacement or cleaning all have an impact on organisations. A supplier's inaccuracy is almost always reported in the media, jeopardising the company's reputation [19]. This affects not just the sales of that product, but also the sales of many other products provided by the company's warehouse or retailers, and the products may be automatically forbidden or rejected. The ramifications of such regulations on the food manufacturing industry are varied, complex, and extensive [20].

In China, forty to sixty percent of buyers refused to buy native milk products due to melamine contamination, while those who bought imported milk powder increased from thirty-four percent to forty-seven percent [21].

Impacts on Farmers

Adulteration affects not only large corporations, but also farmers and producers (of dairy, honey, coffee, wheat, and other products), who are harmed by the weakest link in the industry chain. Many farmers incurred considerable losses during the crisis, as well as cost increasesdue to feed expenses, milk cow shortages caused by mass sales or slaughter, as in the case of the China dairy scandal, and a lack of acceptability of their products [22].

Impact on Consumers

Adulteration of food has a significant negative influence on human health. Diarrhoea, abdominal discomfort, nausea, vomiting, headaches, cancer, anaemia, insomnia, muscular paralysis, brain damage, and stomach diseases such as giddiness, joint pain, and liver damage have all been connected to food adulteration, as shown in Table 3 [23]. Controlling counterfeit and substandard goods and services comes with a lot of difficulties. For detecting fraudulent and taught products, there is a scarcity of knowledge and technology [24]. The fundamental issue with food adulteration, according to Asrat and Yilma, is market acceptance due to scepticism about its authenticity [25].

Table 3. Adulterants and their adverse effects

Adulterant	Adverse Effects
Water (in milk)	Diluted milk cause malnutrition and impure water maycause intestinal problems.
Starch (in ghee)	May cause diarrhoea and is fatal to diabetic patients.
Urea (sugar)	Failure of kidneys.
Detergent (in milk)	Gastro-intestinal and kidney issues.
Formalin (in milk and milk products)	Poisonous fatal effect.
Chlorine (in packaged drinking water)	Can cause clogging in the arteries and develop heartcomplications. Disturbs acid base balance of body andtampers blood pH.

Source: V. Lakshmi., 2012. R.V. LABS IJSIT, Volume 1, Issue 2, 106. [23]

Challenges of Food Adulteration

Because of numerous important food safety incidents in recent years, there has been a substantial amount of media attention, public health impact, as well as a decline in customer confidence. Consumers, industry, lawmakers, and regulatory agencies have all increased their focus on food safety as a result of this. Inadequate laws, money, and staffing appear to be common difficulties to the management of inferior and counterfeited goods and services, not only in developed but also in developing countries. Furthermore, there are few means to detect false and contaminated items, and there is insufficient technology or knowledge to do so [26].

The precise quantity of foreign suppliers of products is frequently unclear, and only a small fraction of facilities is covered in terms of physical site surveillance. Food adulteration has a fundamental problem in that it is not accepted in the market due to a lack of trust in its originality or a depressed market. Dairy goods from Wolaita and Kucha, for example, were noted for their exceptional quality and consequently were costly. Wolaita butter was later revealed to be tainted with vegetable oil and sold for a low price in the towns of Dilla and Shashemene [27].

Such practises should be avoided since they can tarnish a region's reputation for high-quality butter. As a result, rules and regulations should be in place to punish criminal dealers and dishonest manufacturers that threaten the uniqueness of distinct food products in various locations. Other problems include the need for more selective sample preparation and testing technology, which should be adopted as quickly as feasible to enable speedy screening of food adulteration, as well as effective analytical procedures to detect fakes [28].

It's no surprise that a few major food safety incidents in recent years drew a lot of media attention, had a public health impact, and lowered consumer confidence. As a result of this, consumers, the food industry, policymakers, and regulatory authorities have all grown more aware of the need to protect meals [29].

If they are not satisfied, criminals in underdeveloped nations cannot be prosecuted as a life-threatening threat to social control authorities. Furthermore, there is a scarcity of data and technology to track counterfeit and soiled items. Normally, the specific origin of a product from another country is unclear, but physical website inspections revealed only a fraction of the factory [30]. As a result, restrictions are needed to impose penalties on illegal traders and fraudulent producers who infect completely different foods in various locations, and effective analytical procedures to change the fast-food fake screen have been discovered. It necessitates a great deal of careful sample preparation and testing methodologies [31].

Probable Intervention

Food Safety and Standards Act of 2006 (34 of 2006) entered into force on August 5, 2011. The Food Safety Act 2006, and its provisions, consolidates the laws related to food and establishes the Food Safety and Standards Authority of India to establish science-based standards for food and to regulate its manufacture, storage, distribution and importation [32].

Under the new law, food business operators, a term that encompasses the 4,444 businesses that handle food in any way, are responsible for providing safe food to consumers.

Among other things, the Act incorporates key provisions of the PFA Act of 1954 and is based on international laws and instruments. In a nutshell, the Act regulates foreign trade practices and envisions a policy that is far-reaching. Those who work in the manufacturing, marketing and distribution industries are subject to regulation. The Indian Food Safety and Standards Authority regulates those who work in the food manufacturing, marketing, processing, handling, transportation, import and selling industries. The Act is current, comprehensive and aims to provide consumers food safety [33].

The Food Safety and Standards Authority of India (FSSAI) has learned of cases of sale/delivery of substandard and adulterated food, including edible oil.

State/UT Food Safety Officials conduct sporadic surveillance, supervision, inspections, and controls of food products, including edible oil, in order to prevent food adulteration, and measures have been taken to ensure compliance with the provisions of the FSS Act 2006 on delinquent food business operators [34].

State food safety authorities have been advised from time to time to maintain strict vigilance by regularly taking samples of food from all sources, i.e., manufacturers, wholesalers, and retailers, and taking strict action against perpetrators in accordance with the provisions of the FSS Act of 2006. This is to ensure the availability of high- quality food for consumers and to keep the problem of food adulteration under control. To review the sale of adulterated edible oils, the FSSAI has notified regulations that require the sale of edible oils only in packaged form, with the exception of the sale of oil in bulk in certain circumstances as permitted by the state government [35].

As part of a central sector initiative aimed at "Strengthening the food analysis system in the country, including the deployment of mobile food analysis laboratories (SoFTeL)," FSSAI has notified 263 primary food testing laboratories and state food analysis laboratories around the country. FSSAI has published a brochure entitled Detect Adulteration with Rapid Test (DART), which covers more than 50 common rapid tests for the detection of food adulteration in the home by citizens themselves in order to raise awareness among citizens. Consumers on food safety. The DART book is available on the FSSAI website. The FSSAI has also introduced Guideline for the Introduction of Rapid Analytical Food Test (RAFT) Kit / Equipment /

Method. States/UTs are principally responsible for enforcing the provisions of the Food Safety and Standards (FSS) Act, 2006, and the Rules and Regulations promulgated there under [36]. Table 4 depicts the cases and convictions, which is a reflection of the legal steps taken in India against the food adulteration practice.

The Global Food Security Act of 2016 was signed into law on July 20, 2016, with significant bipartisan support, marking a watershed moment in US recognition of the crucial role of food security in development and national security. The legislation formalized the United States government's commitment to increasing the productivity, income, and livelihoods of small-scale producers, particularly women, by working across agricultural value chains and increasing farmers' access to local and international markets. It strengthened the existing accountability mechanisms for the Feed the Future initiative and established parameters for rigorous Congressional oversight, monitoring, and evaluation of progress toward this commitment. As a result of the GFSA, the United States has signalled to the international community that food security is a top priority, and it has strengthened monitoring and evaluation of food security programming, developed a more integrated approach, and deepened interagency engagement.

President Trump signed the Global Food Security Reauthorization Act of 2017 into law on October 11, 2018, after it was passed by Congress with overwhelming bipartisan support. This law, which reauthorizes the Global Food Security Act for another five years, sends a strong message that America is committed to addressing the root causes of hunger through Feed the Future [37].

Table 4. The number of cases in last few years

Year	No. of Cases	Conviction	No. of Cases in which Penalties were Imposed	Amount Raised
2016-17	13,080	1,605	4,757	Rs.17,01,93,266
2017-18	15,121	5,198	7,627	Rs.26,35,41,067
2018-19	21,363	701	12,734	Rs.32,57,78,087

Source: "FSSAI Publishes Manual: "Detect Adulteration with Rapid Test (DART)" - Food Safety Helpline" [36].

Conclusion

The amount of dietary misconduct is too large to justify additional exercise to adjust the situation. The method of collecting food samples can lead to biased estimates and therefore can undermine the validity of the crowd. A systematic type of random series of food samples integrated into food fraud monitoring should provide the correct overview of the situation. Appropriate measures are required to make heavy use of various compounds and dyes in tableware. The majority of the customers lack proper knowledge, attitude, and practices concerning food adulteration. Publicizing the newly-surpassed client protection law, extraordinary modern-day food adulteration-related laws, and remarkable additives of food adulteration thru mass media need to play a crucial characteristic in raising client awareness. Stringent enforcement of the upcoming unified food law 'Safe Food Act 2013' through manner of method of the Government may considerably decrease food adulteration within the country.

References

[1] Shaib, S., 2021. *A Feasibility Report on Food Adulteration.*, VIT University.

[2] Anita, G., Neetu, S., 2013. Hazards of new technologies in promoting food adulteration. *J. EnvSci, Tox. F Sci* 5(1), pp. 08-10.

[3] Misgana, B., Food adulteration and some methods of detection, review, *International Journal of Nutrition and Food Sciences.* Vol. 9, No. 3, 2020, pp. 86-94.

[4] Ankita, C., Neeraj, G., Fozia, H., Skarma, C.,2020. An overview of food adulteration: Concept, sources, impact, challenges and detection, review, *International Journal of Chemical Studies*. Vol. 8, Issue 1.

[5] Rakesh, K., 2016. *Methods for Detection of Common Adulterants in Food.*

[6] Food Marketing Technology, 2021. Food adulteration by food marketing technology. *FMT Blogs*.

[7] Coley., Noel., 2005. The fight against food adulteration. *Education in Chemistry*. Vol. 42, no. 2. Royal Society of Chemistry. pp. 46–49.

[8] *Food and Agriculture Organization of the United Nations*, Viale delle Terme di Caracalla, Rome, Italy.

[9] Ali, A., M.S. Mahmood., Iftikhar, H., Masood, A., 2011. Adulteration and microbiological quality of milk (a review). *Pakistan Journal of Nutrition*, 10: 1195-1202.

[10] Asrat, A., Zelalem, Y., Patterns of milk and milk products adulteration in Bodittitown and its surroundings, South Ethiopia. *Scholarly J Agric. Sci.* 2014; 4(10):512-516.
[11] Mehna, S.,2021. *Food Adulteration: Types, Causes, Methods and Detection.*
[12] Sandesh, A., 2018. Food adulteration, types of food adulteration and mitigation measures. *Public Health Notes.*
[13] Ankita, C., Neeraj, G., Fozia, H., Skarma, C., 2020. An overview of food adulteration: Concept, sources, impact, challenges and detection, review, *International Journal of Chemical Studies*. Vol. 8, Issue 1.
[14] Fairchild, G.F., Nichols, J.P., Capps, O., 2003. Observations on economic adulteration of high-value food products: the honey case. *Journal of Food Distribution Research, Food Distribution Research Society*, vol. 34(2), pages 1-8.
[15] Grumezescu, A.M., Holban, M., 2018. *Food Safety and Preservation: Modern Biological Approaches to Improving Consumer Health*. Academic Press.
[16] Misgana, B., *International Journal of Nutrition and Food Sciences,* Ethiopian Institute of Agricultural Research, Jimma Agricultural Research Centre, Jimma, Ethiopia.
[17] Sukumaran, M.K., Navya, P., Raju, K., 2017. analysis of food adulterants in selected food items purchased from local grocery stores., *International Journal of Advances in ScientificResearch* 3(7):82.
[18] Food Safety and Standards Authority of India (FSSAI), 2017. *Detect Adulteration with Rapid Test (DART).*
[19] Pandpal, S.D., Sirvastava, A.K., Negi, K.S., 2012. Estimation of quality of rawmilk (opened and branded) by milk adulteration testing kit. *Ind. J Commu. Health.*; 24(3):188.
[20] Ibens, D.,2014. Impact on enterprises. *Food Quality&Safety*, Current Issue.
[21] Qian, G., Guo, X., Guo, J., Wu, J., 2011. China's dairy crisis : impacts, causes and policy implications for a sustainable dairy industry. *IJSDWE*. 18(5):434-441.
[22] Nie, Y.L., 2008. Impact on farmers. collections of speeches in the summit forum of dairy industry development in China. *China Dairy*. 12:10-19.
[23] Lakshmi, V. 2012. R.V. *LABS IJSIT*, Volume 1, Issue 2. 106.
[24] Chauhan, S., K. D. Mandal., 2019. Adulteration of milk: a review. *International Journal of Chemical Studies*, v5. i6. Pp. 37-41-22.
[25] Asrat A, Zelalem Y. 2014-23. Patterns of milk and milk products adulteration in Bodittitown and its surroundings, South Ethiopia. *Scholarly J Agric. Sci.*
[26] Ankita, C., Neeraj, G., Fozia, H., Skarma, C.,2020. An overview of food adulteration: Concept, sources, impact, challenges and detection, review, *International Journal of Chemical Studies.* Vol. 8, Issue 1.
[27] Yigrem, S., Beyene, F., Tegegne, A., Gebremedhin, B., 2008. *Challenges of food adulteration: Dilla and Shashemene town's dairy production, processing and marketing systems of Shashemene-Dilla area, South Ethiopia.*
[28] Ibens, D., 2014. *Challenges : Food Quality & Safety*, Current Issue.
[29] Ades, G., Henry, C.W., Feldstein, F., 2012. The food safety challenge of the global food supply chain. WEBINARS, *Food Safety Magazine.*

[30] Lawal, B.D., 2012.Overview of the socioeconomic implications and management of product faking and adulteration., *Impact Factor* 2012 (UJRI): 0.6670 ICV 2012: 6.03.

[31] Ibens, D., 2014. The great melamine scares. *Food Quality & Safety*, Current Issue.

[32] *The Food Safety and Standards Act,* 2006.

[33] Jitendra P. D., 2006. *Food Safety and Standards Act*.

[34] Ministry of Health and FamilyWelfare, 2020. Preventing food adulteration. *PIB Delhi* (Release ID: 1604954).

[35] *Food Safety and Standards Act,* 2006a. No. 34 of 2006-33.

[36] FSSAI, 2017. Detect Adulteration with Rapid Test (DART). *Food Safety Helpline*.

[37] *Global Food Security Act*, 2016.

Chapter 6

Isolation of Probiotic Bacteria from Bandel Cheese and Its Application in Food Product Development

Minakshi Chakraborty[1,*], Bedotroyee Chowdhury[1], Debabrata Bera[1] and Lakshimishri Roy[2,†]

[1]Department of Food Technology and Biochemical Engineering, Jadavpur University, Kolkata, India
[2]Department of Food Technology, Techno India, Salt Lake, Kolkata, India

Abstract

Probiotic bacteria are gaining a significant increase in popularity because of their scientifically proven beneficial effect on human gut health. The range of probiotic-containing food products is varied and rampant. Bandel cheese is one of the traditional and obsolete varieties of cottage cheese which originated in the Portuguese colony of Bandel Hooghly, Eastern India. It provides beneficial health effects and contributes some functional properties. In this study, probiotic bacteria were isolated from Bandel cheese. Isolates were identified by biochemical characterization and evaluated for their probiotic property. These isolates showed resistance to low potential hydrogen (pH), tolerance against bile salt, sodium chloride, and temperature. The growth extract of these isolates also showed antimicrobial activity against *Escherichia coli* and *Bacillus subtilus*. In this study, isolated lactic acid bacteria were applied in

* Corresponding Author's Email: minakshichakraborty09@gmail.com.
† Corresponding Author's Email: drlakshmi1371@gmail.com.

In: Proceedings of BIOSPECTRUM
Editors: Sanket J. Joshi, Susmita Mukherjee and Moupriya Nag
ISBN: 978-1-68507-985-7

probiotic yogurt preparation. Relevant compositional parameters, texture profile, pH, and syneresis of yogurt were determined. In addition, the proximate compositional parameters were analyzed to determine the nutritional value of yogurt. The sensory panels of yogurt are rated and accepted concerning color, flavor, mouthfeel, taste, texture. Therefore, in this study probiotic bacteria were isolated from Bandel cheese and were applied in food product development, enhancing the functional property and nutritional value of food.

Keywords: bandel cheese; isolation; probiotic; yogurt; optimization

Introduction

"Probiotic" is a Greek word that means live or active microorganisms [7]. Probiotic incorporated formulated products include foods, drugs, and dietary supplements [7]. A probiotic like *Lactobacillus Sp.* and *Bifidobacterium* are the most popularly used strains for their beneficial effects on human gut health. Lactic acid-producing bacteria i.e., *Lactobacilli* are Gram-positive and Catalase-negative [11]. Lactic acid bacteria produce a protective effect on the human gut flora [6] and are hence used for treating and preventing some infectious diseases.

Bandel cheese is the traditional and obsolete variety of cottage cheese that originated in Bandel [17]. Smoked and Non-smoked are the two varieties available in West Bengal. Presently, the production of Bandel cheese has been shifted from Bandel to other parts of West Bengal [17]. The industrial production of Bandel cheese is yet not standardized. It provides a beneficial health effect and contributes some functional properties [17].

The biodiversity of commercial starters has become limited due to the large-scale use of starter culture for the industrial fermentation process [6]. Non-pasteurized milk and raw materials are still a source of new strains in the case of traditional dairy products [6]. Pure cultures which are isolated from traditionally fermented food products exhibit various metabolic activities that strongly deviate from the ones of comparable strains used as industrial bulk starters of culture [6]. Therefore, an effort should be made to preserve and apply the microbial diversity of traditional Bandel cheese. Hence, this study was initiated to evaluate the probiotic properties of probiotic bacteria isolated from Bandel cheese and their application in the preparation of yogurt at optimal process conditions and eventually determination of the nutritive value of the yogurt.

Methods

Sample Collection

Bandel cheese was collected from the local market, Kolkata, and stored refrigerated in aseptic condition at 4°C for a laboratory experiment.

Isolation of Bacteria

0.5 gm of Bandel cheese sample was dissolved in 10 ml of sterile distilled water. The solution was then thoroughly mixed and serial dilutions from 10^{-1} to 10^{-6} were prepared. 0.1 ml of serially diluted samples were poured on selective MRS (De Man, Rogosa, and Sharpe) agar plates following the pour plate method [2]. Plates were then kept under anaerobic conditions in an incubator at 37°C for 48 hours [5]. Pure isolates were selected from colonies obtained on streak plates. Gram staining characteristics of isolated bacteria were assessed, along with morphological characteristics observed under the microscope [8].

Biochemical Characterization

To differentiate the bacteria Indole, Methyl red, Voges-Proskauer and Citrate test (IMViC test), motility test, Catalase test were performed [11].

Determination of Probiotic Properties

For the determination of probiotic properties, randomly one of the isolates was selected among 7 isolates. Then Acid and Bile salt tolerance, temperature tolerance, salt tolerance [5, 9] were performed.

Antimicrobial Activity of Isolated Bacteria

The selected isolate was inoculated on MRS agar broth and incubated at 37ºC for 48 hours [5]. It was then centrifuged, and the supernatant solution was concentrated in a rotary vacuum evaporator. The antimicrobial activity of concentrated solution was determined against *E. coli* and *Bacillus sp.* [8]. The sensitivity was measured as the diameter of the zone of inhibition surrounding the bore [8].

Application of Isolated Bacteria for Yogurt Preparation

The isolated bacterial strain was applied for the preparation of probiotic yogurt [1]. First, pasteurized cow milk was boiled and then cooled to room temperature. The suspension of bacteria was prepared in sterile distilled water and added to the milk and incubated at different temperatures [3].

Optimization of the Process of Yogurt Preparation

Bacterial suspension (10^8 CFU/ml) was added to the milk (3, 5, 7%), the total solid of milk was 15, 18, 20° Brix maintained. Brix of milk was determined by Brix meter. It was then incubated at 37ºC, 40ºC, 42ºC temperature for 10 to 12 hours [3, 10].

Determination of Physicochemical, Proximate Composition, and Sensory Analysis

Physicochemical and proximate property of yogurt was determined by standard AOAC (Association of Official Agricultural Chemists) 2012, 2005, 1990 method [10, 14, 15, 16]. Sensory analysis of Yogurt was done using a 9-point hedonic scale [3, 14, 15]. Ten panelists, regular yogurt eaters tasted and described the color, flavor, texture, taste, slime, and the whole formation of yogurt [3, 14, 15].

Determination of Texture Property of Yogurt

Textural characterization of freshly prepared yogurt was evaluated using a texture analyzer instrument (Stable Micro Systems) with an A/BE disc 35 back extrusion probe [4, 13].

Results

Probiotic bacteria were isolated using selective MRS media at 37°C temperature under anaerobic conditions [1, 11]. Out of 68 isolates, seven isolates were selected and characterized, shown in Table 1. Out of seven isolates, one isolate, JULABm (Jadavpur University laboratory m strain) selected for cell morphology, biochemical test, probiotic characteristics, antimicrobial property, and these isolates were applied in yogurt preparation [1, 5]. Isolates were Gram-positive, rod shape [11] shown in Figure 1. The characteristics and morphology of 7 different isolated colonies have been explained in Table 1. The colony of JULABm isolate is circular, flat, and whitish, mentioned in Table 1. IMViC tests indicated isolates are anaerobic or facultatively anaerobic, as catalase-negative [11] shown in Table 2.

Table 1. Colony morphology of seven isolates

Isolates	Colony Shape	Colony Edge	Colony Elevation	Colony Color
JULAB a	Circular	Irregular	Flat	White
JULAB b	Circular	Irregular	Flat	Creamy
JULAB c	Circular	Flat	Convex	White
JULAB m	Circular	Flat	Convex	White
JULAB p	Circular	Flat	Convex	Creamy
JULAB q	Circular	Wavy	Flat	Creamy
JULAB d	Circular	Wavy	Flat	White

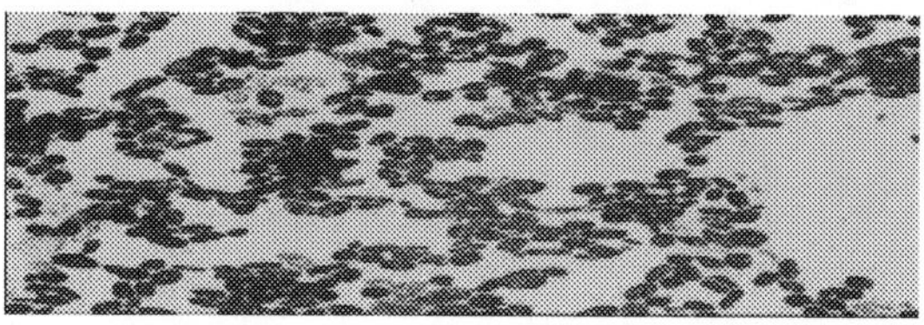

Figure 1. Microscopic view of julabm indicating gram-positive rod shape bacteria.

Table 2. Biochemical characterization of JULABm isolates

Isolates	Gram's Nature	Catalase	Methyl-Red	Voges-Proskauer	Indole	Citrate
JULABm	Positive	Negative	Positive	Negative	Negative	Negative

Table 3. Probiotic property of JULABm isolate

Property	Tolerance
Salt	2-3%
pH	2.5-7.5
Bile salt	0.3 -0.5%
Temperature	36°C-40°C

Probiotic Property of JULABm Isolate

Probiotic properties of isolated bacteria were tested that isolate can tolerate low pH, Bile salt, and temperature up to 40°C [8, 9, 10] shown in Table 3.

Antimicrobial Activity of JULABm Isolates

An assay of antimicrobial activity of JULABm against two pathogenic bacteria showed that the isolated bacteria was capable to inhibit the pathogen to some extent [8] as shown in Figure 2. Isolates JULABm showed a relatively high inhibition zone against *Bacillus Sp.* (26 mm) and *E. coli* (20 mm).

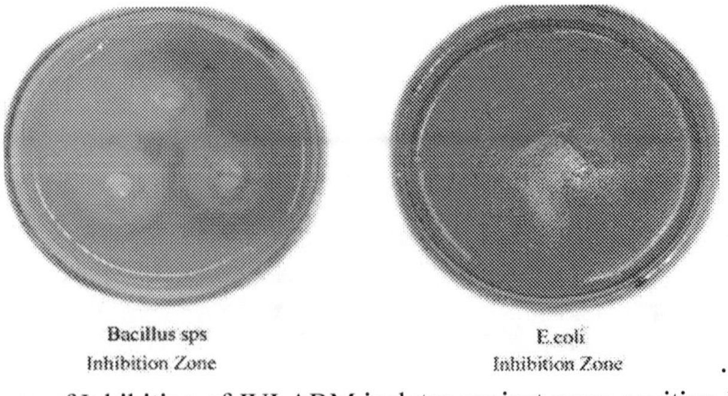

Figure 2. Zone of Inhibition of JULABM isolates against gram-positive *Bacillus sp.* and gram-negative *E. coli*.

Application of JULABm Isolates for Yogurt Preparations

As the isolates showed some probiotic properties, it was then applied for the preparation of probiotic yogurt [1] shown in Figure 3. Optimization of the yogurt preparation, followed by some steps. 3, 5, 7 percent of bacterial suspension was added to 15, 18, 20, and 25° Brix of milk and incubated for 12 to 15 hours at 37°C, 40°C, and 42°C temperature and compared with control [2]. Control was prepared by yogurt culture available in the market. It observed that 3 to 5% bacterial suspension, 18 to 20° Brix, 37°C to 40°C temperature for 12 to 15 hours of incubation is ideal for yogurt preparation. As the temperature increased by 40°C curd became coagulated or separated and as the Brix decreased to less than 18, it increased syneresis.

Similarly, as the ° Brix increased by 20 to 25 the hardness of yogurt increased, compared with control. So, in this study yogurt was prepared with 3, 5, 7% inoculum in 18° Brix at 37°C temperature for further evaluation.

Figure 3. Application of JULABm isolates in yogurt preparation.
This figure represented yogurt prepared with 3 to 5% inoculum, 18°Brix.

Physicochemical and Proximate Composition

The results of physicochemical and proximate composition of yogurt of 18° Brix in different inoculum percentages are shown in Figure 4.

The results of sensory analysis of yogurt (Brix °18) with different inoculum and control are shown in Figure 5.

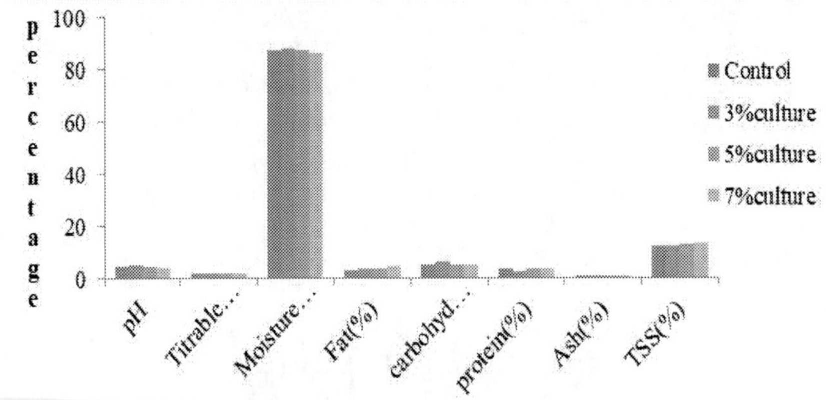

Figure 4. Indication that the yogurt prepared with isolated JULABm isolate was closely like the control. It contained Moisture 86.7 to 87.9%, Fat 3.5 to 4.4%, Protein 2.4 to 3.5%, Carbohydrate 5.3 to 6.1%, and 12 to 13.4% Total Soluble Solid (T.S.S), pH varies from 3.8 to 4.9, and Titratable Acidity 2.01 to 2.11.

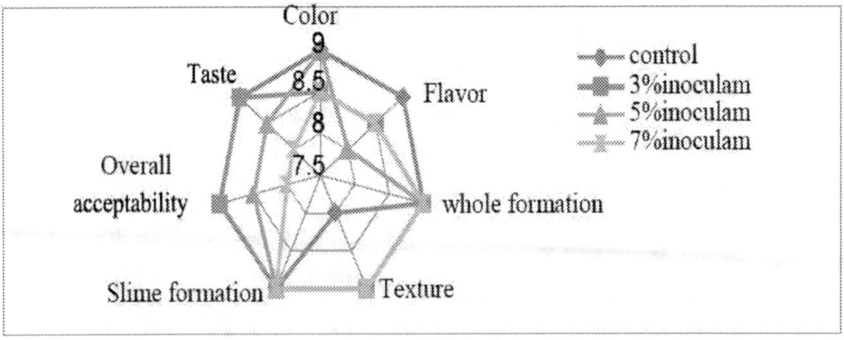

Figure 5. Indication that sensory properties of yogurt prepared with 3-5% inoculum rated more acceptable concerning color, flavor, texture, and overall acceptability when compared with control.

Texture Profile Analysis Attribute of Yogurt of 18° Brix in Different Inoculum Percentage

The texture attributes estimated by the texture profile analyzer [8, 13] are shown in Figure 6. Results indicated that the hardness of yogurt prepared with 7% inoculum represented a little higher than the control. Similarly, adhesiveness increased as the inoculum's percentage increased. The yogurt prepared with 3% and 5% inoculum showed similar results compared with control concerning chewiness, gumminess, resilience [8, 13].

Figure 6. Indication of texture profile analysis of yogurt, using texture profile analyzer stable micro system A/BE disc 35 back extrusion probes.

Table 4. TPA attribute of yogurt of 18° Brix in different inoculums percentage

TPA Attributes of yogurt (18° Brix)	Control	3% inoculum	5% inoculum	7% inoculum
Hardness	225	203	237	270
Adhesiveness	-2023	-1959	-2008	-2103
Springiness	0.934	0.984	0.9325	0.923
Cohesiveness	0.514	0.317	0.476	0.493
Gumminess	112.89	102.5	120.7	134.19
Chewiness	108.0	62.72	105.14	122.86
Resilience	0.009	0.009	0.008	0.007

Discussion

Bandel cheese is an obsolete variety of cottage cheese prepared by cow milk [17]. Milk is an excellent source of lactic acid bacteria [12]. Therefore, in our study, MRS Agar media was used to enhance the growth of lactic acid bacteria as MRS is selective media for *Lactobacillus* sp. [1, 8, 9]. JULABm strain was selected among the variety of colonies of 7 isolates, as m strain showed a clear colonial zone [8], shown in Table 1.

Isolates are not decomposed hydrogen peroxide into oxygen, as they are not utilized in catalases. Isolates are citrate negative since isolates are the sole source of carbon, absence of blackening of media are indicating the absence of Hydrogen Sulphide production, these findings are similar to the literature [11], mentioned in Table 2.

Isolated JULABm strains showed positive results in gram staining. The microscopic view is shown in Figure 1 confirms the presence of a thick peptidoglycan layer outer the cell membrane, that retains a crystal violet stain and makes a strong bond with the polypeptide layer, which indicates isolates are gram-positive bacteria [11].

Determining the probiotic property acid tolerance and bile tolerance are crucial parameters [2, 8, 9]. As probiotics are supplemented orally, they must have the ability to survive in the gastrointestinal tract [9]. Tolerating acidic conditions indicate that isolates can withstand gastric stress and survive high acidic load of food without affecting the host [2, 8, 9].

Bile tolerances indicate isolates can enter the small intestine and maintain the intestinal gut flora [8, 9, 12]. Bile salts are secreted from the upper part of the gastrointestinal tract of healthy humans. Hence, for effective use of probiotic bacteria, the species must be tolerant to 0.3% to 0.5% of bile salt [2, 12]. A study indicating, JULABm isolates were able to survive 0.3% bile concentration is shown in Table 3.

Simultaneously, our JULABm isolates are capable to inhibit the pathogen to some extent [8]. Lactic acid bacteria are capable to produce organic substances, antimicrobial metabolites during growth [9]. The isolated strain can also produce bacteriocin against gram-positive and gram-negative pathogenic bacteria, shown in Figure 2. Compared with the literature [8] isolates showed a relatively high inhibition zone against Ecoli (20 mm) and bacillus (26 mm) [8], shown in Figure 2.

Probiotics bacteria are generally recommended as safe bacteria [9]. Therefore, our isolates are applied in yogurt preparation (shown in Figure 3) to determine the effect of supplementation of probiotic bacteria [1, 2]. In this study, the process of yogurt preparation was also optimized [2]. Physiochemical and sensory parameters of yogurt were also compared with the literature [10], indicating that the yogurt prepared with isolated JULABm strains was closely similar to the literature [10]. The percentage of moisture is 86.7 to 87.9%, fat 3.5 to 4.4%, protein 2.4 to 3.5%, carbohydrate 5.3 to 6.1%, and 12 to 13.4% of total soluble solids, pH varies from 3.8 to 4.9, and titratable acidity 2.01 to 2.11[10] in yogurt. The study also indicated 3 to 5% inoculum is acceptable for yogurt preparation. As the inoculum percentage increases, the acidity of yogurt also increased, which decreases the overall acceptability of yogurt shown in Figure 5.

Different studies have proven that texture, appearance, and functional property are greatly influenced by the structural arrangement of the protein network [4]. The Texture Profile Analysis is mentioned in Table 4. Hardness indicated how much force is required to attain a certain deformation [4, 13]. Springiness property indicated the ability to recover after the first decomposition [4]. Gumminess is the result of hardness and cohesiveness if hardness increases the gumminess also increased. Chewiness is the product of springiness, hardness, and cohesiveness, it refers to the energy required to masticate or chew the food [4, 13]. Resiliency indicates recovering back to the original shape. Cohesiveness indicates the ability of binding or to hold together. When our yogurt is compared with the literature [4], similar results are given in hardness, gumminess, but differ in cohesiveness and springiness, resiliency, therefore further texture optimization is required.

This indicates that the isolated bacteria produce antibacterial metabolites like organic acid or bacteriocin during growth. Considering that our isolated bacteria were used for the preparation of probiotic yogurt and the production process of yogurt was optimized. This study indicated that 3 to 5% inoculum is ideal for yogurt preparation.

Conclusion

Bandel cheese is a unique and fermented dairy product. It has ethnic importance among West Bengal. Preparation of Bandel cheese is an art of technology passed on by generations. This study focused on the identification of phenotype characteristics of lactic acid bacteria. Identification of genotype characteristics is still necessary. Isolated lactic acid bacteria traits showed antimicrobial activity. It produces lactic acid bacteriocin. Bacteriocin is potentially used in the food industry to control contaminant bacteria, but the availability of bacteriocin is limited and expensive. Therefore, finding Lactic acid bacteriocin producers from local and traditional Bandel cheese is cost-effective and necessary.

Different external factors, environmental pollution, sedentary lifestyle, poor eating habits, smoking, anxiety alter the human gut microflora that triggers various gastrointestinal disorders like diarrhea, constipation, indigestion, and lactose intolerance. Probiotics help to balance the gut microflora and improve the digestibility of food. Therefore, the application of probiotic bacteria in yogurt increases the functional property and nutritional value. However, to enhance quality, symbiotic yogurt can also be prepared by supplementing prebiotics like inulin, oligo-fructose, etc with probiotics.

Acknowledgments

The authors are thankful to the Department of Science and Technology, Government of West Bengal for their financial support.

References

[1] Choi, H. S., Lim, S. Y., 2019, Viability of Probiotic Bacteria in Yogurt Supplemented with Enzyme - Bio Converted Ginseng, Ascorbic acid, and Yeast Extract. *Korean Society of Milk and Biotechnology*, 37(1), 57-68.

[2] Hyronimus, B., Marrec, C. L., Sassi, A. H., Deschamps, A., 2000, Acid and Bile Tolerance of Spore-forming Lactic Acid Bacteria. *International Journal of Food Microbiology*, 61, 193-197.

[3] Khandelwal, P., Upendra, R. S., Amiri, R. Z., 2014, Optimization of the Process Parameters for the Preparation of Synbiotic Yogurt-Cheese. *International Journal of Pharma Sciences and Research,* 5, 980-987.

[4] Kose, Y. E., Altun, I., Kose, S., 2018, Determination of Texture Profile Analysis of Yogurt Produced by Industrial and Traditional Method. *International Journal of Scientific and Technological Research,* 4, 66-68.

[5] Luiz, L. M. P., Castro, R. D., Sandes, S. H. C., Silva, J. G., Oliveira, L. G., Sales, G. A., Nunes, A. C., Souza, M. R., 2016, Isolation and Identification of Lactic Acid Bacteria from Brazilian Minas Artisanal Cheese. *Journal of Food,* 15:1, 125-128.

[6] Prasant., Tomar, S. K., Singh, R., Gupta, S. C., Arora, D. K., Joshi, B. K., Kumar, D., 2009, Phenotype and Genotype Characterization of Lactobacilli from Churpi Cheese. *Dairy Science Technology*, 89, 531-540.

[7] Salam, M. A., 2014, Probiotic Concepts and Application. *Bangladesh Journal of Medical Science*, 13, 373-77.

[8] Sari, M., Suryano, D., Yurnaliza., 2018, Antimicrobial Activity of Lactic Acid Bacteria Isolated from Bekasam against *Staphylococcus aureus* ATCC 25923, *Escherichia coli* ATCC 25922 and *Salmonella* sp. *Earth and Environmental Science,* 130, 1-8.

[9] Shehata, M. G., El Sohaimy, S. A., El shan, M. A., Youssef, M. M., 2016, Screening of Isolated Potential Probiotic Lactic Acid Bacteria for Cholesterol-Lowering Property and Bile salt Hydrolase Activity. *Annals of Agricultural Science,* 61(1), 65-75.

[10] Tagne, T. A., Kamda, S. A., Ponka, R., Fokou, E., 2021, Physicochemical Evaluation and Microbiological Properties of Artisanal Yogurt Sold in Maroua-Cameroon. *Food Science and Nutrition Technology,* 6(1).

[11] Thakur, M., Deshpande, W. H., Bhate, A. M., 2017, Isolation and Identification of Lactic Acid Bacteria and Their Exploration in a Non-dairy Probiotic Drink. *International Journal of Current Microbiology and Applied Science,* 6, 1023-1030.

[12] Vantasawa, A. P., Maryah, T. U., Timothy, B., 2017, Isolation and Identification of Lactic Acid Bacteria with Probiotic Potential from Fermented Cow Milk (Nono) in Unguwar Rimi Kaduna State Nigeria. *American Journal of Molecular Biology,* 7, 99-106.

[13] Zheng, Y., Liu, Z., Mo, B., 2015, Texture Profile Analysis of Sliced Cheese in Relation to Chemical Composition and Storage Temperature. *Journal of Chemistry,* 2016, 1-10.

[14] A.O.A.C. 1990, *Official Methods of Analysis of the Association of Analytical Chemists*. 15th ed., Arthington, U.S.A.

[15] Association of Official Agricultural Chemists- AOAC.2005, *Official Method of Analysis of AOAC International*. 18th ed. Maryland. U.

[16] A.O.A.C. 2012, *Official Methods of Analysis of the Association official Analytical Chemistry International* 19th ed., Association of Analytical Communities, Gaithersburg.

[17] Dutta. R., 2013, All You Wanted to Know about Local Cheese- Kalimpong to Bandel. *The Telegraph,* Calcutta, India. May 3, Retrieved 19 January.

Chapter 7

Characterization of Black Gram Nuggets (*Bori*): Traditional Product of Midnapore District, West Bengal, India

Najmun Nahar, Utpal Raychaudhuri and Sunita Adhikari (Nee Pramanik)*

Department of Food Technology and Biochemical Engineering, Jadavpur University, Kolkata, West Bengal India

Abstract

Black gram nuggets (bori), a popular traditional pulse-based food item in India, are prepared from aerated paste (batter) of soaked black gram (*Phaseolus mungo*) and have high nutritional importance. In this study, samples of spinach bori, carrot bori, tomato bori, wax-gourd bori, and daikon bori were collected. Plain black gram bori was used as a control. Carbohydrate, protein, fat, ash content along with moisture, antioxidant content, and antioxidant activity were analyzed. The incorporation of vegetables improved the antioxidant content and antioxidant activity of bori samples. Improvement in total phenolic content to the range of 0.76-1.15 mg/g, total flavonoid content to the range 0.77-1.33 μg/g was observed. FRAP assay activity to the range of 6.92-11.48 μmol/g, and DPPH assay activity to the range of 7.50-36.04%, against control was observed. Total carbohydrate content and fat content also improved to the range of 0.84-2.39% and 8.00-15.50% respectively. Protein and ash content were decreased to the range of 1.35% - 2.82% and 1.11-10.31% against control. The moisture content of all samples increased during storage compared to the initial day. Moisture content helped to indicate

---

* Corresponding Author's Email: sunitapramanik@gmail.com.

In: Proceedings of BIOSPECTRUM
Editors: Sanket J. Joshi, Susmita Mukherjee and Moupriya Nag
ISBN: 978-1-68507-985-7

their shelf life. Principal component analysis justifies the correlation of characterization parameters and bori varieties. Texture profile analyzer helps to analyze their textural properties.

Keywords: nuggets, nutrition, antioxidant, texture, Principal Component Analysis (PCA)

Introduction

Legumes are a valuable source of protein and other nutrients. Black gram (*Phaseolus mungo*) is widely consumed for protein, fiber, vitamins, and minerals. It is also a good source of antioxidants like polyphenols, flavonoids, phytosterols that prevent several diseases.[1] Common names of black gram are urad dal (Hindi), mashkalai dal (Bengali), biri dal (Odia), adad (Gujaratis), manha di dal (Punjabi), uluntu (Tamil), and kalo dal (Nepalis), etc. Mashkalai dal belongs to the Leguminosae family. Although it is highly consumed all over Asia and Europe, very few studies have been performed on the nutritional benefits of these legumes.[2] Dehusked cotyledon of black gram is used in many Indian dishes, especially in fermented dishes such as idli, dosa, and non-fermented foods like hopper, papad, waries (spicy hollow balls), and *bori*.

Nugget is a pulse-based popular traditional food item of the Midnapore district, West Bengal in India. It is mainly prepared in rural areas by women using mashkalai dal during winter. Overnight soaked pulses are first ground to a smooth paste. It is then aerated to attain proper consistency and density, and this form is known as a batter.

Small aliquots of batter of approximately 10 g weight are manually poured on mustard oil to rub over the tray, clean cloth, or wooden stage in arrays and sun-dried for two to three days until crispy nuggets with hollow inner core are obtained.[3] Nuggets are consumed after oil frying or as an added item in a variety of Indian dishes. It has a good demand in several parts of the country.

---

[1] Sreerama, Sashikala, and Pratape, "Variability in the Distribution of Phenolic Compounds in Milled Fractions of Chickpea and Horse Gram: Evaluation of Their Antioxidant Properties."

[2] Bhattacharya, Latha, and Bhat, "Controlled Stress Rheological Measurement of Blackgram Flour Dispersions."

[3] Swami, Das, and Maiti, "Convective Hot Air Drying and Quality Characteristics of Bori: A Traditional Indian Nugget Prepared from Black Gram Pulse Batter."

Some micro-entrepreneurship projects have been developed around this product.

Vegetables are a rich source of vitamins, minerals, and antioxidants.[4] Tomatoes, carrots, and spinach are rich sources of antioxidants, vitamins like lycopene, beta carotene, vitamin-A, vitamin-C, vitamin K, folate, and minerals like potassium, iron, calcium, etc.[5] Radish and wax-gourd are rich sources of polyphenol compounds, isothiocyanate compounds like sulforaphane, alpha-lipoic acid, vitamin-C, dietary fiber, iron, vitamin-B1, phosphorus, zinc, etc.[6] These vegetables can decrease the possibility of cancer, diabetes mellitus, heart disease, and promote wellness.[7] Vegetables are incorporated in the *bori* primarily to enhance their nutritional and antioxidant capacity.

To our best knowledge and belief, this study has not been done before. The objectives of the study were to analyze the nutritional characteristics like carbohydrate, protein, fat, etc., antioxidant content and activity, and textural properties of different *bori* specimens. These *bori* specimens are traditional products of the Midnapore district, West Bengal, India.

Methodology

Sample Collection

Six varieties of nuggets (bori), black gram bori (BB), spinach bori (SB), carrot bori (CB), wax gourd bori (WB), tomato bori (TB), and daikon bori (DB) were collected from several blocks in East and West Midnapore and were stored in an airtight container at room temperature for further study.

---

[4] Kaur and Kapoor, "Anti-Oxidant Activity and Total Phenolic Content of Some Asian Vegetables."

[5] Sharma et al., "Chemical Composition, Functional Properties and Processing of Carrot—a Review"; Bhowmik et al., "Tomato-A Natural Medicine and Its Health Benefits"; Tewani, Kumar Sharma, and Rao, "Spinach (Palak) Natural Laxative."

[6] Jurica and Petříková, "Nutritional and Sensory Value of Conventionally vs Organically Grown Chinese Radish (Raphanus Sativus L. Var. Longipinnatus)"; Abdullah et al., "Study on Antioxidant Capacity and Phenolic Content of Various Parts of Wax Gourd (Benincasa Hispida)."

[7] Bhowmik et al., "Tomato-A Natural Medicine and Its Health Benefits."

Antioxidant Content and Antioxidant Activity

Extract Preparation

1g sample was taken and mixed with 20 mL of 80% methanol (HPLC grade) in a 100 mL Erlenmeyer flask. After sonicating in an ultrasonic bath (D-150/IH, Shanti Industrial Estate, India) for 20 min, it was centrifuged at 8900 rpm for 10 min at 4 °C. The supernatant was subjected to subsequent analysis for antioxidant content and antioxidant activity.

Total phenol content (TPC), total flavonoid content (TFC), DPPH radical scavenging capacity, and Ferric reducing antioxidant power (FRAP) were determined following the procedure described by Ray et al.[8]. TPC was expressed as mg of gallic acid equivalents (GAE) per g dry weight (DW). TFC was measured as mg of quercetin equivalent (QE) per g DW of samples. DPPH free radical scavenging capacity was expressed in total antioxidant activity.

$$Total\ antioxidant\ activity = \frac{Blank - Sample}{Blank} \times 100 \quad (1)$$

The reducing power of the extracts was determined by ferric reducing antioxidant power (FRAP) expressed as µmol of ascorbic acid equivalent (AAE) per mg of extract. The antioxidant activity was calculated from a calibration curve prepared using a standard concentration of ascorbic acid.

Nutritional Analysis

The total Carbohydrate content of the samples was measured by the method of Akindahunsi and Vetayo.[9] 0.5 g of all *bori* powder was taken in a 250 ml digestion flask and 50 ml of 2.5 N HCl solution was added to it. Hydrolysis was done by boiling the solution in a water bath for three hours at 70 °C. After cooling, it was filtered using Whatman filter paper, then it was neutralized by NaOH solution, and pH was adjusted to 7. The supernatant was stored for analysis. Glucose was used as standard. 4 ml anthrone reagent was added to

---

[8] Ray et al., "Preparation of Okra-Incorporated Dhokla and Subsequent Analysis of Nutrition, Antioxidant, Color, Moisture and Sensory Profile."

[9] Akindahunsi and Oyetayo, "Nutrient and Antinutrient Distribution of Edible Mushroom, Pleurotus Tuber-Regium (Fries) Singer."

all samples, followed by heating for eight min in a water bath. After cooling, absorbance was measured by UV-Visible spectrophotometer at 630 nm.

$$Carbohydrate\ (\%) = \frac{weight\ of\ glucose\ (g)}{weight\ of\ sample\ (g)} \times 100 \quad (2)$$

The protein content of the sample was determined by using the Folin Lowry method.[10] 0.5 g of all sample flour was extracted in 10 ml of 0.1 M phosphate buffer (pH - 6.5) in a rotary shaker for one hour. Then the solution was centrifuged at 1500 rpm for 30 min and the supernatant was used for protein estimation. Here, bovine serum albumin was used as standard. 1 ml distilled water was used as blank in a test tube. 4.5 ml reagent I (0.1 N NaOH added in 48 ml 2% Na_2CO_3 in water + 1 ml of 1% NaK tartarate in water + 1 ml 0.5% $CuSO_4,5H_2O$ in water) was taken in all sample test tubes, then incubated for 10 min. After incubation, 0.5 ml of reagent II (1:1: Folin phenol: water) was added. This was followed by incubation for 30 min. After that, absorbance was measured by UV-Visible spectrophotometer at 660 nm.

$$Protein\ (\%) = \frac{weight\ of\ albumin\ (g)}{weight\ of\ sample\ (g)} \times 100 \quad (3)$$

Fat, moisture and ash content was measured by the method AOAC.[11] 5 g of all samples were taken in thimbles and sealed with cotton, then transferred to Soxhlet apparatus and 250 ml petroleum ether was filled into a bottle and then placed over a mantle heater. The process is continued for 6 hr at 60-70°C. After completing the process, the flask was completely dried, placed in a desiccator, and reweighed.

$$Fat\ (\%) = \frac{weight\ of\ the\ fat\ (g)}{weight\ of\ sample\ (g)} \times 100 \quad (4)$$

5 g sample was taken in a crucible and placed in a muffle furnace at 550°C for 5 hrs, till the ash was formed. After this, the weight of ash was measured.

$$Ash\ (\%) = \frac{weight\ of\ ash\ (g)}{weight\ of\ the\ sample\ (g)} \times 100 \quad (5)$$

[10] Lowry et al., "Protein Measurement with the Folin Phenol Reagent."
[11] AOAC: "Official Methods of Analysis."

3 g sample was taken in a metal dish and put in a hot air oven at 105 ℃ for 3 hr until a constant weight was achieved. Then the sample dish was placed in a desiccator for 10 min to cool down the temperature. After that, the weight of the sample was measured.

$$\textit{Moisture content} = \frac{\textit{Initial weight of sample} - \textit{final weight of sample}}{\textit{Initial weight of sample}} \times 100 \quad (6)$$

Samples were stored at room temperature and moisture content was observed on the first day as well as after 15 days, 1 month, and 2 months of storage.

Textural Profile Analysis (TPA)

TPA of *bori* Nugget was performed using a TA.XT. Plus texture analyzer (Stable Micro System Ltd., UK), following the procedure described by Swami et al.,[12] A 36 mm diameter (P/36R) aluminum cylindrical probe was used. The nugget was placed centrally under the compression plate and TPA was performed five times. The compression distance was 30% of the nugget height. Test speed was 1 mm/s, pretest and post-test speed were 1 mm/s, time pause was 5s. TPA tests were carried out using a TA. XT texture analyzer (Stable Micro System Ltd., UK, model TA. XT Express). The analyzer was linked to a computer that recorded data via a software program called Exponent Light Express (Stable Micro System Ltd., UK) which generated a plot of force (grams) versus time (s). A 36 mm diameter (P/36R) aluminum cylindrical probe was used to measure the TPA of *bori* samples.

Statistical Analysis

All values were presented in triplicate (n = 3) and data was presented as mean and standard deviation. The statistical analysis of results were done by the R.(3.4.3) (R development Core Team, 2017, Vienna, Austria) for principal component analysis (PCA).

---

[12] Swami, Das, and Maiti, "Texture Profile Analysis of Cooked Sun Dried Nuggets (Bori) Prepared with Different Levels of Moisture Content and Percent Air Incorporation in Its Batter."

Results and Discussion

Antioxidant Content and Antioxidant Activity

From Table 1, it was observed that incorporation of vegetables in bori improved the TPC content to the range of 0.76-1.15 mg/g and TFC content to the range of 0.77-1.33 µg/g compared to plain bori. FRAP activity too was improved to the range of 6.92-11.48 µmol/g. However, in the case of DPPH, all samples showed a marked increase in free radical scavenging activity around 7.50 to 36.04%. SB is high in TPC (1.15 µg/g), FRAP (11.48 µmol/g), and DPPH (23.93%) as compared to other vegetable bori. DB has a high TFC (1.33 µg/g) value, as compared to the other vegetable nuggets.

Table 1. Antioxidant, carbohydrate, protein, fat, and ash content analysis of all *bori* samples. The control *bori was* designated as black gram *bori* (BB), other vegetable *bori* were designated as spinach *bori* (SB), carrot *bori* (CB), wax gourd *bori* (WB), tomato *bori* (TB), and daikon *bori* (DB). The values present the mean of independent experiments ± SD

ID	TPC	TFC	FRAP	DPPH (%)	Carbohydrate (%)	Protein (%)	Fat (%)	Ash (%)
BB	0.54 ± 0.010	0.55 ± 0.030	3.04 ± 0.005	17.59 ± 0.010	59.30 ± 2.300	23.03 ± 1.398	2.00 ± 0.050	9.89 ± 0.900
SB	1.15 ± 0.050	1.01 ± 0.020	11.48 ± 0.500	23.93 ± 0.020	57.32 ± 1.690	24.93 ± 0.498	2.25±0.059	10.10 ± 0.700
CB	0.76 ± 0.029	0.95 ± 0.010	9.46 ± 0.300	23.50 ± 0.010	60.20 ± 1.149	22.38 ± 0.798	2.00±0.010	8.90 ± 0.800
TB	0.96 ± 0.019	1.17 ± 0.050	8.28 ± 0.010	20.86 ± 0.050	60.20 ± 1.513	23.21 ± 0.598	2.16±0.020	9.01 ± 0.700
WB	1.03 ± 0.010	0.77 ± 0.100	10.39 ± 0.010	18.91 ± 0.010	59.80 ± 0.987	22.63 ± 0.980	2.31±0.010	8.87 ± 0.500
DB	0.99 ± 0.010	1.33 ± 0.050	6.92 ± 0.030	23.72 ± 0.040	60.72 ± 1.199	22.72 ± 0.750	1.99±0.019	9.78 ± 0.398

Nutrition

From Table 1, it was observed that incorporation of vegetables in *bori* improved their nutritional value such as carbohydrate, protein, and fat content. Total carbohydrate content was improved to the range of 0.84-2.39%, but in the case of SB, carbohydrate content was decreased by 3.34% compared to BB. Being a leafy vegetable spinach, had a very low level of carbohydrate content (2.9 g /100 g) but was a rich source of protein and minerals compared to other vegetables.[13] The protein content of the SB sample increased to 8.25% and the TB sample increased to 0.78% compared to BB. Similarly, other sample's value decreased to the range of 1.35-2.82% compared to BB. Fat content was improved too for all samples, from about 8.00 to 15.50%. Ash content of all samples was decreased to about 10.31-1.11% compared to BB. Ash content improvement (2.12%) was observed in SB as compared to control. DB had higher carbohydrate content than the other vegetable nuggets. SB had a higher protein and ash content than other *bori.* Fat content was higher in WB than in other nugget samples.

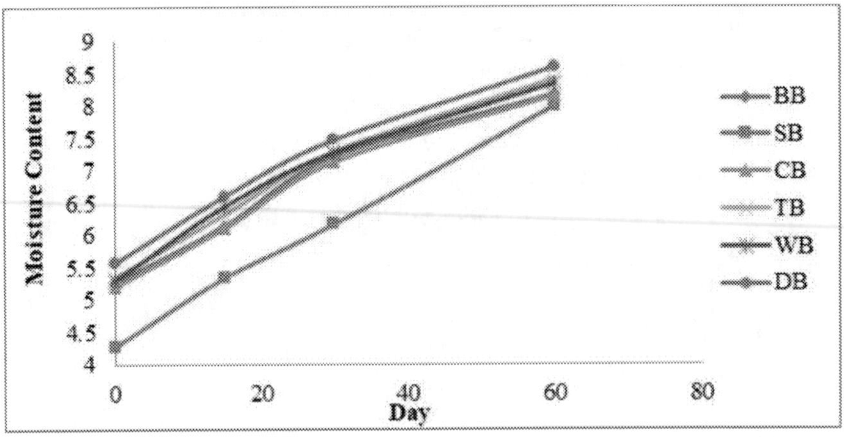

Figure 1. Variation of moisture content of the different bori samples with storage time. The control bori was designated as black gram bori (BB), other vegetable bori were designated as spinach bori (SB), carrot bori (CB), wax gourd bori (WB), tomato bori (TB), and daikon bori (DB). The values were presented as the mean of independent experiments ± SD.

Moisture Content

Quality, as well as storage characteristics of a food product, are dependent on moisture content since the growth of spoilage organism was favored by high

[13] Gopalan, Rama Sastri, and Balasubramanian, *Nutrition Value of Indian Foods.*

moisture content.[14] It was observed that for sample BB, the initial moisture content was 5.26%. After 15 days of storage the moisture content was increased to 16.54% to the initial day. After 60 days, the moisture content of BB was 55.51%. All samples had shown increased moisture content after 60 days of storage, and the rate of water uptake was also increased for all samples except SB. SB samples took water at a greater rate, i.e., 82.67% (60 days) compared to the initial day.

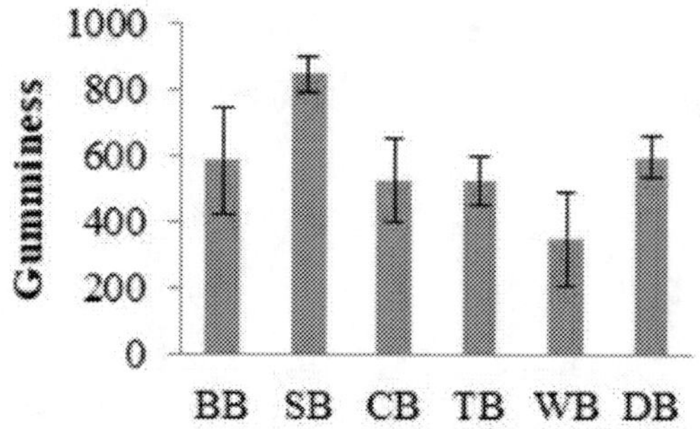

Figure 2 (a)

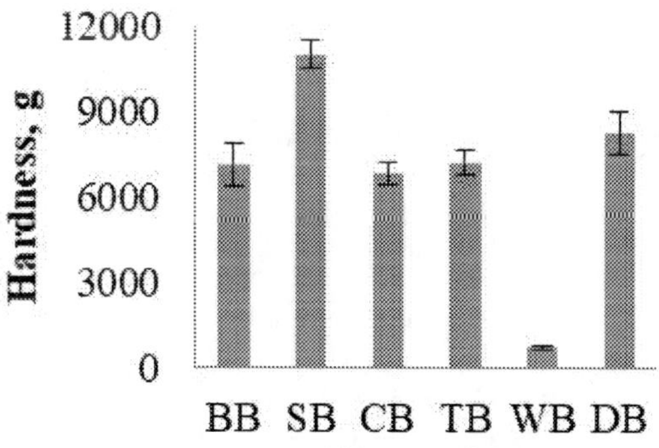

Figure 2 (b)

Figure 2. (Continued on next page).

[14] Swami, Das, and Maiti, "Moisture Sorption Isotherms of Black Gram Nuggets (Bori) at Varied Temperatures."

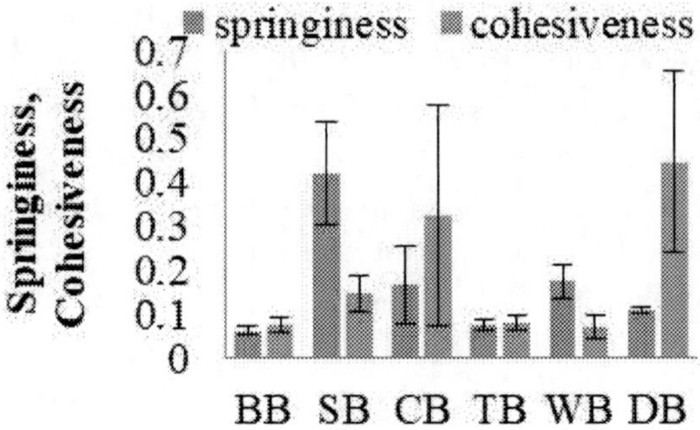

Figure 2 (c)

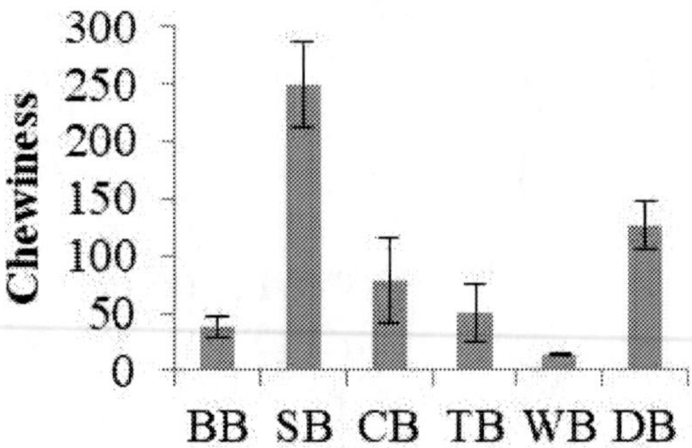

Figure 2 (d)

Figure 2. (a, b, c, d) Analysis of variance for gumminess, hardness, springiness, cohesiveness, and chewiness of different *bori* samples. The control *bori was* designated as black gram *bori* (BB), other vegetable *bori* were designated as spinach *bori* (SB), carrot *bori* (CB), wax gourd *bori* (WB), tomato *bori* (TB), and daikon *bori* (DB). The values were presented as the mean of independent experiments ± SE.

Texture Profile Analysis (TPA)

Figure 2 (a) shows the gumminess of the different varieties of *bori* samples. Gumminess values were increased for SB and DB and decreased for CB, TB, and WB against control. The gumminess for the SB sample was high and WB was low for other samples. Figure 2 (b) showed the hardness of the nugget samples. Hardness increased in SB, TB, and DB but decreases in WB and CB compared to BB (control). Figure 2 (c) shows the springiness and cohesiveness of all samples. Springiness and cohesiveness values were increased for all samples compared to control, but cohesiveness was decreased only for WB. Figure 2 (d) showed the chewiness of all black gram nuggets. Chewiness values were increased in all samples except WB as compared to control. SB (Spinach *bori*) had high gumminess, hardness, chewiness, and springiness values to other samples. Cohesiveness values were high in DB compared to others. WB had low gumminess, hardness, chewiness, cohesiveness values compared to other samples, but springiness was high in TB. In all parameter studies of texture profiles, the TB and BB samples were similar.

Principal Component Analysis (PCA)

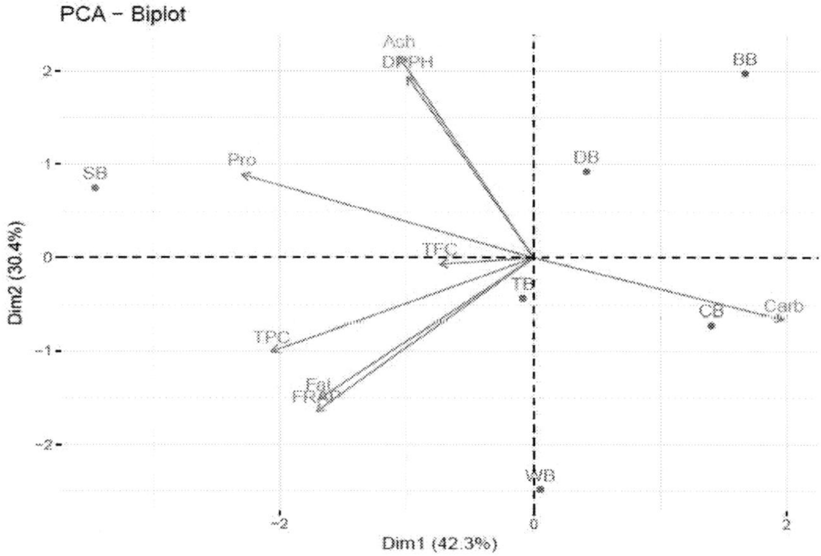

Figure 3. PCA – Bi-plot of all samples with their parameters. The control bori was designated as black gram bori (BB), other vegetable bori were designated as spinach bori (SB), carrot bori (CB), wax gourd bori (WB), tomato bori (TB), and daikon bori (DB). The values were presented as the mean of independent experiments ± SD.

It was concluded from the analysis of the six principal components (PC) produced and the percentage of variation observed that the first three PC would be adequate for explaining the variance caused by different ingredients.

PC1 showed a positive correlation with carbohydrates, while PC2 showed a positive correlation with protein, ash, DPPH and a negative correlation with others. From the Bi-plot, it could be seen that BB, DB, TB, and CB samples were similar with high carbohydrate content and lower fat, protein, ash, antioxidant value.

In contrast, SB had high protein, fat, ash, and antioxidant values. WB sample was characterized by high TPC, FRAP, fat and carbohydrate values, and lower protein, ash, DPPH activity.

Conclusion

All *bori* had a significant impact on nutrition profile, antioxidant properties, and moisture content. All vegetable *bori* showed improvement in their antioxidant content and antioxidant activity (TPC content: 0.76-1.15 mg/g, TFC content: 0.77-1.33 µg/g, FRAP activity: 6.92-11.48 µmol/g and DPPH activity 7.50-36.04%) compared to BB. SB had the highest TPC content (1.15 mg/g), DPPH (36.04%), and FRAP activity (11.48 µmol/g), whereas, WB contained more TFC (1.33 µg/g) than all other samples.

Carbohydrate and fat content were improved to the range of 0.84-2.39% and 8.00-15.50% respectively but in the case of SB, carbohydrate content was decreased to 3.34% compared to BB used as a control. SB was good for nutrition. It was a rich source of protein (8.25%), fat (15.50%), and ash (2.12%) but a poor source of carbohydrate. Its moisture absorption was higher than the other samples, so its shelf life was very low. Ash content of all samples was decreased to the range of about 1.11-10.31% compared to BB.

The principal component analysis confirmed the findings of the characterization parameters. With the help of texture profile analysis, it was concluded that SB showed high values for all parameters except cohesiveness; similarly, WB had low values for all parameters except springiness. Cohesiveness was high in DB. The correlation of characterization parameters was done by PCA. From this study, we can conclude that the incorporation of specific vegetables in *bori* preparation improved their antioxidant content and nutritional values.

Acknowledgments

This research work was financially supported by the UGC-MANF Scheme, Government of India.

References

Abdullah, N, S W K W Saidatul, Z Samicho, K S Zulkifli, and N Aziman. "Study on Antioxidant Capacity and Phenolic Content of Various Parts of Wax Gourd (Benincasa Hispida)." *World Applied Sciences Journal* 19, no. 7 (2012): 1051–56. https://doi.org/10.5829/idosi.wasj.2012.19.07.2900.

Akindahunsi, A A, and F L Oyetayo. "Nutrient and Antinutrient Distribution of Edible Mushroom, Pleurotus Tuber-Regium (Fries) Singer." *LWT - Food Science and Technology* 39, no. 5 (2006): 548–53. https://doi.org/10.1016/j.lwt.2005.04.005.

AOAC. *Official Methods of Analysis,* 1, 2005. https://www.researchgate.net/publication/292783651_AOAC_2005.

Bhattacharya, S, R B Latha, and KK Bhat. "Controlled Stress Rheological Measurement of Blackgram Flour Dispersions." *Journal of Food Engineering* 63, no. 2 (2004): 135–39. https://doi.org/10.1016/S0260-8774(03)00291-7.

Bhowmik, D, K P S Kumar, S Paswan, and S Srivastava. "Tomato-A Natural Medicine and Its Health Benefits." *Phytojournal 1*, no. 1 (2012): 33–43. https://www.phytojournal.com/archives/2012/vol1issue1/PartA/3.pdf.

Gopalan, C, B V R Sastri, and SC Balasubramanian. *Nutrition Value of Indian Foods.* (1980). https://agris.fao.org/agris-search/search.do?recordID= US201300551774.

Jurica, M, and K Petříková. "Nutritional and Sensory Value of Conventionally vs Organically Grown Chinese Radish (Raphanus Sativus L. Var. Longipinnatus)." *Horticultural Science* 41, no. 2 (2014): 64–70. https://doi.org/10.17221/200/ 2013-hortsci.

Kaur, C, and H C Kapoor. "Anti-Oxidant Activity and Total Phenolic Content of Some Asian Vegetables." *International Journal of Food Science and Technology* 37, no. 2 (2002): 153–61. https://doi.org/10.1046/j.1365-2621.2002.00552.x.

Lowry, O H, N J Rosebrough, AL Farr, and RJ Randall. "Protein Measurement with the Folin Phenol Reagent." *The Journal of Biological Chemistry* 193, no. 1 (1951): 265–75. https://doi.org/10.1016/s0021-9258(19)52451-6.

Ray, S, SK Saha, U Raychaudhuri, and R Chakraborty. "Preparation of Okra-Incorporated Dhokla and Subsequent Analysis of Nutrition, Antioxidant, Color, Moisture, and Sensory Profile." *Journal of Food Measurement and Characterization* 11, no. 2 (2017): 639–50. https://doi.org/10.1007/s11694-016-9433-x.

Sharma, K D, S Karki, N S Thakur, and S Attri. "Chemical Composition, Functional Properties and Processing of Carrot—a Review." *Journal of Food Science and Technology* 49, no. 1 (2011): 22–32. https://doi.org/10.1007/S13197-011-0310-7.

Sreerama, Y N, VB Sashikala, and VM Pratape. "Variability in the Distribution of Phenolic Compounds in Milled Fractions of Chickpea and Horse Gram : Evaluation of Their Antioxidant Properties." *Journal of Agricultural and Food Chemistry* 58 (2010): 8322–30. https://doi.org/10.1021/jf101335r.

Swami, S B, S K Das, and B Maiti. "Moisture Sorption Isotherms of Black Gram Nuggets (Bori) at Varied Temperatures." *Journal of Food Engineering* 67, no. 4 (2005): 477–82. https://doi.org/10.1016/J.JFOODENG.2004.05.014.

Swami, S B, S K Das, and B Maiti. "Convective Hot Air Drying and Quality Characteristics of Bori: A Traditional Indian Nugget Prepared from Black Gram Pulse Batter." *Journal of Food Engineering* 79, no. 1 (2007): 225–33. https://doi.org/10.1016/j.jfood eng.2006.01.064.

Swami, S B, SK Das, and B Maiti. "Texture Profile Analysis of Cooked Sun-Dried Nuggets (Bori) Prepared with Different Levels of Moisture Content and Percent Air Incorporation in Its Batter." *International Journal of Food Engineering* 3, no. 5 (2007). https://doi.org/10.2202/1556-3758.1155.

Tewani, R, J K Sharma, and S V Rao. " Spinach (Palak) Natural Laxative." *International Journal of Applied Research And Technology* 1, no. 2 (2016). http://www.ijart.info/.

Chapter 8

Bacteriocins: A New Group of Antimicrobials for Food Preservation

Nishantika Mitra[1], Ankita Dey[2], Moupriya Nag[3,*], Dibyajit Lahiri[3], Vijay Jagdish Upadhye[4] and Rina Rani Ray[2,†]

[1]Manipal Academy of Higher Education, Manipal, India
[2]Department of Biotechnology,
Maulana Abul Kalam Azad University of Technology,
Haringhata, West Bengal, India
[3]Departement of Biotechnology,
University of Engineering & Management,
Kolkata, West Bengal, India
[4]Center of Research for Development (CR4D),
Parul University,Vadodara, Gujarat, India

Abstract

Food spoilage and wastage is an alarming crisis globally emphasizing urgent measures to be taken for the preservation of raw food and food products. Although chemical preservatives and antibiotics are conventionally used for food preservation, the toxic effects of the former and the development of antibiotic resistance due to the use of later are increasingly refraining people from using these chemicals. One plausible strategy for preventing food spoilage involves the usage of small bacterial peptides known as bacteriocin having bactericidal effects brought about either through the disintegration of the cell wall or

* Corresponding Author's Email: moupriya.nag@uem.edu.in.
† Corresponding Author's Email: raypumicro@gmail.com.

In: Proceedings of BIOSPECTRUM
Editors: Sanket J. Joshi, Susmita Mukherjee and Moupriya Nag
ISBN: 978-1-68507-985-7

inhibition of DNA and protein synthesis of the food rotting bacteria. Bacteriocins are ribosomally synthesized, hydrophobic anti-microbial peptides that can inhibit food spoilage caused by both Gram-positive and Gram-negative bacteria. Although these are found to be produced by both Gram-positive and negative bacteria, the lactic acid bacteria (LAB) are noteworthy for their ability to synthesize various types of bacteriocins. Unlike antibiotics, these are primary metabolites of bacteria, tolerant to high thermal stress, and remain active at a wide pH range. Out of all, Nisin has particularly gained more attention as it is regarded safe for application in the food industry and is widely used for the preservation of meat and dairy products. Bacteriocins possess the ability to induce post-translational modification and can be easily degraded by proteolytic enzymes in the human gastrointestinal system making them safe for human consumption. Being primary metabolites, they can be easily ameliorated through bioengineering which will increase their bioavailability as well their target specificity.

Keywords: bacteriocin, antimicrobial resistance, nisin, food preservation

Introduction

Spoilage of food and consequent food loss has been the biggest threat to food security and extensive search for addressing this urgent problem is to be prioritized. Since the etiological agents for different food spoilage of which some are fatal and debilitating [1, 2], suitable agents are to be used to check such spoilage causing organisms. The food destroying pathogens like *Clostridium botulinum* [1] *Listeria monocytogenes* [3], *Enterococcus* [4], *Streptococcus, Staphylococcus, Bacillus,* and *Escherichia coli* produce various types of toxins and after invading human tissues can cause inflammatory diseases, respiratory infections, systemic infections and intestinal disorder to cancer. To increase the shelf life of food products, several chemical preservatives are used, but most of them are detrimental to human health. Hence the demand for the use of bio preservatives or natural preservatives is rising [5]. The lactic acid bacteria (LAB) are found to possess a high potential to be used as bio preservatives as they can produce active antimicrobial substances like organic acids, bacteriocins, hydrogen peroxide, etc. [6].

A large number of bacterial cells comprising of both Gram-positive and Gram-negative species synthesizes proteins possessing antimicrobial activities known as bacteriocins that possess the ability to inhibit or destroy the growth of other microorganisms [7]. Various studies suggested that almost all bacterial species are responsible for synthesizing bacteriocins [8-9]. They are the groups of ribosomally synthesized peptides that possess antimicrobial activity against closely related strains to the producer (narrow spectrum) and sometimes they also act against a wide range of organisms across genera (broad-spectrum bioactivity [10]. Bacteriocins are usually the group of low molecular weight (usually less than 10 kDa) peptides that undergo the mechanism of post-translational modifications and can be easily degraded in the presence of the proteolytic enzymes, especially the proteases being produced by the mammalian gastrointestinal tract thus making it safe for human consumption. They are the group of amphipathic, cationic molecules as they are made up of the excess amount of arginyl and lysyl residues [11-12]. They are usually unstructured when present in an aqueous solution but become structured when exposed to organic solvents like trifluoroethanol or mixed anionic phospholipid membranes [13].

Table 1. Difference between bacteriocin and antibiotics [29]

Bacteriocins	Antibiotics
1. Applied clinically and in food industries	Applied only in clinical practice
2. Ribozomally synthesised	Synthesised as secondary metabolites
3. Its activity is against a narrow bacterial spectrum	Its activity is against a broad bacterial spectrum
4. It has a high proteolytic enzyme degradability	It has medium to moderate proteolytic enzyme degradability
5. Have a high thermal stability	Have a low thermal stability
6. it has a wide active pH range	It has a narrow pH range
7. In terms of taste, colour and odour bacteriocins shows nothing making them applicable to be used a natural food preservative	The antibiotics have a wide range in taste, colour and odour.
8. These are flexible to be applied in bioengineering	These are not flexible to be applied in bioengineering

Among the various groups of Gram-positive bacterial cells, Lactic acid bacteria (LAB) have gained particular importance due to their ability to produce bacteriocin [14-17]. LAB is predominantly used in dairy, meat, bakery, and vegetable fermentation from time immemorial [18-20]. LABs possess the ability to produce various types of antimicrobial substances that include diacetyl acetoin, organic acids, hydrogen peroxide, reutericyclin, reuterin, antifungal peptides, and bacteriocins [21-23]. Bacteriocins act as the most effective natural food preservatives [24] and also contribute to preserving the fermented products [24-26]. The bacteriocins being produced by the LAB have gained a lot of importance from the industries for their various importance and potential applications. The bacteriocins have been widely applied in food packaging as they possess the antimicrobial film that helps in the preservation of the food for a longer period [27]. Though the bacteriocins show properties similar to that of antibiotics, it differ much from conventional antibiotics in various aspects (Table 1) [28].

Experimental

Table 2. Selective list of bacteriocin isolated from microorganisms and can be used as food preservatives

Sl. No.	Microorganism Producing Bacteriocin	Source	Reference
1	Plantaricin LPL-1, from *Lactobacillus plantarum* LPL-1	Fermented fish	[30]
2	Bacteriocin from *Lactobacillus murinus* AU06	Marine sediments	[31]
3	*Enterococcus faecium* NKR-5-3/Enterocins NKR-5-3A, B, C, D, Z	Fermented fish	[32]
4	*Enterococcus faecium* m3a/Enterocin A, P, B	Rabbit meat	[33]
5	*Lactococcus lactis* subsp. *Lactis* 69/ Nisin Z	Beef jerky	[34]
6	*Lactobacillus plantarum* BM-1/Plantaricin BM-1	Fermented meat	[35]
7	*Lactobacillus alimentarius* FM-MM$_4$/ Lactocine MM$_4$	Fermented meat product	[36]
8	*Lactobacillus plantarum* DY4-2/DY4-2	Fish	[37]
9	Bacteriocin Lac-B23, from *Lactobacillus plantarum J23*	Chinese traditional fermented milk	[38]
10	Bacteriocin from *Leuconostoc lactis* SM2	Fermented dairy products	[39]

Bacteriocin possesses excellent antimicrobial activity against food-borne pathogens and is increasingly being used as an alternative to chemical preservatives. Many researchers have focussed on the efficient ways of isolation and purification of novel bacteriocins to be used as food preservatives (Table 2).

Classification of Bacteriocins

Over the years various scientists provided the concepts of classification LAB bacteriocins [40]. The type of classification that was given by Cotter and co-workers (2005) has been the most acceptable among the scientific communities [41]. LAB-producing bacteriocins are small, heat-stable, amphiphilic, and membrane permeabilizing agents. These LAB bacteriocins can be majorly classified into 3 classes. The anionic cell wall of the bacteriocins possesses lipoteichoic and teichoic acids which play an important role in the initial interaction of these anionic bacteriocins. Moreover, these bacteriocins have a higher range of antibacterial or antimicrobial activity at a lower pH as the properties and the cell wall of the bacteriocin are pH-dependent. Coming to the 3 major classifications of the LAB bacteriocins, are Lantibiotics, non-Lantibiotics, and Bacteriocins [42].

Lantibiotics

These are heat-stable peptides that are made up of polycyclic thioether amino acids which are methyllanthionine or lanthionine. These are also composed of unsaturated amino acids that are 2-aminoisobutyric acid and dehydroalanine [43]. These classes of bacteriocins are characterized by their post-translational modification; Nisin and Lactocin are their representatives. These are extensively post- transitionally modified resulting in the formation of unusual amino acids like lanthionine and methyllanthionine. This class of bacteriocin is heat stable peptides that become functionally active after post-translational modification. They can be divided into two types out of which type A are relatively elongated, screw-shaped, and are flexible molecules. Nisin and Lacticin 3147 are the two main representor of this category [44].

Non-Lantibiotics

These classes of LAB bacteriocins are also small, heat-stable molecules but are non-lanthionine-containing peptides [45]. They can also be divided into four categories: namely class II a, (pediocin like bacteriocin), II b (peptide-like bacteriocin), II c (circular bacteriocin), II d (unmodified, linear, non-pediocin like bacteriocins). The class IIa bacteriocin has a specific conserved sequence in their N- terminal region which gives them a strong special inhibitory activity against highly pathogenic microorganisms namely Listeria monocytogenes. Also, this class of bacteriocin is called the pediocin because the first bacteriocin to be discovered under this group was PA-1. While the class IIb bacteriocins are comprised of two peptides whose functioning synergistically is essential for the activity of the bacteriocins [46]. Studies showed that this characteristic of class IIb bacteriocins is responsible for their antimicrobial activity which is influenced by the synergistic functioning of the peptides [47]. The class IIc bacteriocin is not clearly understood rather poorly understood but they are discussed according to its circular configuration structure. Also, their N and C terminal are covalently attached thereby resulting in a very strong molecular structure and a strong anti-microbial action as well [48]. Class IId that usually comprise of one peptide and is of type non-pediocin [9].

Bacteriocins

These are group heat-stable like other classes however these have a larger molecular weight as compared to others [45] (Table 3).

Table 3. Classification of bacteriocins

Class	Charateristics
Class I	The bacteriocins are post-translationally modified, linear or globular peptides, small in size, low molecular weight and are composed of lanthionine and 3-methyllanthionine.
Class II	Heterogeneous class of small peptides known as non-lathionine containing peptides.
II a	Small, heat labile in nature. Synthesized in the precursor form and shows anti-microbial activity against Listeria monocytogenes. It also contains sequence of YGNGNXC in its N-terminal region.
II b	Contains two separate peptides. The activation of its anti-microbial activity requires the functioning of the two peptides together.
II c	It has a circular structure as the N and the C terminal are covalently linked to each other.
II d	Consists of the remaining group of bacteriocin which can be defined as one peptide, non-pediocin miscellaneous type.
Class III	These are comparatively larger molecules and are heat unstable proteins. Also, these peptides are sensitive in nature.

Genes Responsible for Bacteriocin Synthesis

The performance of bacteriocin containing N-Terminal leader sequence is encoded by the structural bacteriocin gene which helps prevent the activation of bacteriocin when present within the producer cell and helps in the recognition of the transporter system. The leader glycine sequence comprises 14-30 residues [49]. The consensus elements are present within the double glycine leader comprise of two glycine residues present at the C-terminus of the cleavage site, conserved hydrophobic and hydrophilic residues that remain separated by defined conserved residues. The double glycine leader comprises of minimum 14 amino acids. Among the consensus residues being present within the leader only the glycine at the -2 positions remain fully conserved. The mature bacteriocins are determined by a length of 30 residues but not more than 100 residues. It has been observed that colicin V comprises 80 residues that are being produced by E. coli which is formally grouped under class II bacteriocin [50-51]. Usually, bacteriocins comprise double peptides both of which comprise double glycine leader and contain contiguous genes within the same operon. The peptides cannot be structurally differentiated from one another and both are required for the optimal activity of the bacteriocin. The antimicrobial efficacy of lactococcin MN and lactococcin G is completely dependent upon the peptides being present at the leader sequences [52].

The bacteriocins under class II also share some common features like possessing high contents of amino acid-like glycine. They are highly cationic possessing a pI within the range of 8 to 11 and comprise of the hydrophobic domain and amphiphilic region that provide its action upon the membranes [53]. The activity of Lacticin F is regulated by two functional genes *tafA* and *lafX* that enhance the ability of its inhibitory effect on other organisms [54].

Biosynthesis of Bacteriocins

Bacteriocins possess simple biosynthetic machinery as they are primary metabolites that are encoded by genes and synthesized by ribosomes when compared to commercial antibiotics [55]. The bacteriocins are biologically inactive precursor peptides that contain an N terminal leader peptide which is then attached to a C terminal proteopeptide. As a result, the inactive bacteriocin undergoes enzymatic inactivation a process referred to as bacteriocin maturation where they are transformed into mature active bacteriocins [56]. In this process of biosynthesis, the leader peptide has the following roles, maturation process and transport of bacteriocin to extracellular space, recognize sites for the biosynthetic enzymes; bacteriocin being present within the producer cell the leader peptide protects the producer strain from the inhibitory activity of the bacteriocin, therefore keeping the bacteriocin inactive while it is inside, 3. The enzyme-substrate interaction provides a suitable confirmation [56]. The biosynthesis of bacteriocin involves the producer strain that is the ribosome synthesizes the Nisin A which is the inactive peptide. The modification enzymes present named Nis B and Nis C involve dehydration and cyclize the propeptide. This is followed by the mechanism of translocation of modified precursor which is responsible for the cleavage of leader peptide thus releases the active Nisin A which is the mature form. Moreover, the Nis I lipoprotein protects the mature bacteriocin from its anti-microbial action by binding to the Nisin A molecules around the producer cells.

Bactericidal Properties of Bacteriocin

The antibacterial activity of the bacteriocins is class-specific. The cationic nature of the bacteriocin bestows them the advantage of the initial interaction with the anionic cell membrane of the target bacterial strain. The electrostatic interaction between the bacterial cell membrane (negative charge) and the bacteriocin molecule (positive charge) is the root cause of the bacterial killing.

Class I Bacteriocin or Lantibiotics: Killing Action

Since the cell wall composition of the Gram-negative and the Gram-positive bacteria are different the former being composed of lipopolysaccharide outer membrane and the LAB bacteriocin is particularly active against the Gram-negative bacteriocin provided that certain surfactants are present which aids in killing as it bears some integrity of the outer membrane [57]. The mechanism of killing by bacteriocin requires a receptor or a docking molecule in the target bacterial strain which differs from different classes of bacteriocin, ions, peptides, amino acids which ultimately results in the killing.

Since Lipid II, the central component in bacterial cell wall synthesis is a target for antimicrobial, Nisin binds with the lipid leading to the blocking of the transportation of peptidoglycan from the cytoplasm to the cell wall and thereby stops the synthesis of cell wall leading to the death of the bacterial cell.

Accumulation of a higher concentration of the Nisin-Lipid complex leads to pore formation finally resulting in bacterial cell death. Thus, Nisin is proved to be very efficient in the killing of pathogenic microorganisms [58].

Class II Bacteriocin or Non-Lantibiotics: Killing Action

These are the most common group of bacteriocins. These bacteriocins execute their antimicrobial action by membrane permeabilization which causes leaking of cytoplasmic molecules. However, the mechanism of this may differ from other classes of bacteriocins.

Class II A Bacteriocin (Pediocin Bacteriocin): Killing Action

The class II a bacteriocins known as the pediocin like bacteriocin that has a potent action toward the highly pathogenic food-borne bacteria named Listeria monocytogenes. These express their anti-microbial action by attacking the sugar uptake system like the mannose phosphotransferase system of the target microorganisms by acting as docking molecules resulting in bacterial cell death. The N-terminal region of the class IIa bacteriocin is responsible for the anti-listerial antimicrobial activity whereas, on the other hand, the C-terminal region of the class IIa Lantibiotics plays the role of an antimicrobial agent against the target pathogenic organisms [59].

Class II B Bacteriocin (Pediocin Bacteriocin): Killing Action

This class of bacteriocin consists of two peptide subunits that work together synergistically to exert the antimicrobial property of this bacteriocin against the target bacterial strain. When tested individually these peptides have a very low anti-microbial action. Therefore, the two peptide subunits are considered together as an antimicrobial agent rather than two individual units [60]. This class can be further divided into type E and type S [61] their anti-bacterial action is shown by membrane permeabilization of the bacterial cell membrane resulting in leaking of the cytoplasmic molecules of the bacterial cell. The pore formation action of this class is weaker and the size of the pore is smaller when compared with the Lantibiotics [62].

Class II C Bacteriocin (Pediocin Bacteriocin): Killing Action

As compared to the other classes of bacteriocin this class II c bacteriocin has a wider range of antimicrobial activity over food-borne spoilage and pathogenic bacteria as well. However, the mode of its anti-microbial action is similar to other classes of bacteriocin in the sense that these also exhibit membrane permeabilization as a mode of killing resulting in dissipation or leaking of cytoplasmic molecules of the bacterial pathogen and ultimately death [63]. The fact where it differs is that these do not require a receptor molecule or a docking molecule for its action. Similarly, it had been thought that the basic amino acid units present of the hydrophobic globular structure are involved in the non-receptor intracellular killing but later in a research study conducted by, concluded that gravicin Ml another circular member of the bacteriocin family, in them the presence of ABC transporter protein leads to the efflux of intracellular solutes which ultimately leads to the death of the bacterial cell [63].

Mechanisms of Action of Bacteriocin against Food Spoilage Bacteria

There are several different mechanisms of bacteriocin action by which it exhibits its antimicrobial action to kill a pathogen or to prevent food spoilage. Among these mechanisms, the PMF Depletion mechanism, the Membrane Insertion & Pore Formation mechanism, and the Genetic control mechanism play an effective role in enhancing their activity or function.

Depletion of PMF (Proton Motive Force) in Food Spoilage Bacteria

As stated by research studies every microorganism bears an ATPase in its plasma membrane. This membrane-bound ATPase plays a crucial role in both the aerobic and anaerobic energy transduction pathways. Or in other words, it can be stated that the ATPase catalyzes the ATP synthesis during the oxidative phosphorylation and also helps in the coupling of the ATP hydrolysis to the essential membranes which constitute events involving energy generation or metabolism [64]. These observations were supported by the chemiosmotic theory of Mitchel. According to Harold evidence present in the anaerobic functioning of the ATPase has been observed when studying the anaerobic *S. faecium*. It has been observed that these establish a pH gradient as well membrane potential and that led to the study of the membranes of *E. coli* which lead to the establishment that ATP synthesis is associated with the PMF that is Proton Motive Force. Therefore, according to studies established in microorganisms the ATP synthesis is catalysed only when an inward force that is an inward proton Motive Force is present across the cell membrane [65]. The role of the bacteriocins as observed in the research studies states that there are bacteriocins from the lactic acid bacteria namely, Pediocin PA -1, Leuconocin S, and Lactacin F. Therefore, as observed these lactic acid bacteriocins plays their antibacterial role by depletion of the PMF insensitive microorganisms. As per research the Pediocin PA -1 and Leuconocin S are responsible for the dissipation of the majority of *Listeria monocytogenes* whereas the Lactacin F plays a role in the 87% depletion of the microorganism *Lactobacillus delbrueckii*. Though the Pediocin PA -1 and Leuconocin S are energy-independent Nicin is established to be energy-dependent. Thus, it has been the LAB bacteriocins share a similar process of depletion of PMF as depletion of PMF will inhibit several energy-demanding processes involved in bacterial membrane including ATP synthesis [66]. As a result of depletion, it will lead to the death of the harmful pathogen therefore bacteriocin exhibits its antimicrobial or anti-bacterial properties through this mechanism.

Membrane Insertion and Pore Formation in Food Spoilage Bacteria

Listeria monocytogenes, Bacillus cereus is a major group of pathogenic bacteria causing bacterial infections and food spoilage. Therefore, in such case antibiotics plays a very important role in preventing these food spoilages. But the demands of natural antibiotics of decrease in the chemical preservatives which are present in the commercial antibiotics have led to the development of natural antibiotics or bacteriocins. Among such is Pentocin MQ1. The purification involved the absorption-desorption of the bacteriocin. This type of bacteriocin exhibited a wide spectrum of antimicrobial activity and has a high chemical and thermal activity and has a more stable pH and it is effective against the pathogens such as *Listeria monocytogenes, Bacillus cereus* which causes food spoilage and bacterial infection. One of the researches studies has shown the application of this bacteriocin Pentocin MQ1 benefited by increasing the shelf life of a banana. The main interest of concern is the mechanism by which this bacteriocin works. The mechanism involves cell membrane permeabilization or membrane insertion by pore formation.

In one of the research studies experiments were done to understand the mechanism of pore formation. In this experiment, *M. luteus* was grown in a Mueller Hilton Broth under certain air pressure and SYNTOX green dye was added to it. Following this certain microliter of stained bacteria was added to it and until a stable baseline was reached, 10 microliters of Pentocin MQ1was added to it. Then the fluorescence was observed as the SYNTOX green leaked into intracellular DNA (Consequent death of the bacteria) and was monitored with the help of Real-Time PCR. As per the results Pentocin MQ1produced a higher intensity of fluorescence which indicated the pore formation and therefore membrane permeabilization, however, Nisin produced a higher intensity of fluorescence [67].

Therefore, according to, the LAB bacteriocins exhibited their anti-microbial properties mainly by pore formation and by inhibiting cell wall biosynthesis. For some cases, high intensity of the fluorescence was observed within 30 minutes but for some cases, it took a long time, however, the fact is that nisin adapts the pore formation mechanism to kill pathogens and it is a more rapid killing mechanism. Similar research studies based on LAB Bacteriocins such as the Nisin and the class 1 bacteriocin that is Lantibiotics and their mechanism of membrane insertion and pore formation has been expressed. According to LAB bacteriocins contains these above-mentioned pore-forming peptides. These bacteriocins usually have a wide range of activity and they bear a stable pore, however, class 2 bacteriocins being small and heat-stable have a narrow range of activities. As mentioned, these bacteriocins cause membrane interaction and pore formation with the anionic lipid bilayer that is present abundantly in the membrane of the Gram-positive bacteria. Their activities are mainly enhanced by the presence of docking molecules and receptor that makes the pore stable. While the Lantibiotics produce wedge-like nodes, similarly class 2 bacteriocins produce barrel stave pore with the help of carpet mechanism.

There is a certain step that goes along with pore formation. These are the proper steps that occur before a stable pore is established:

Membrane Interaction in Food Spoilage Bacteria

Gram-positive bacteria are known for their abundance of anionic lipid that is present in their membrane. Because of the cationic nature of the bacteriocin, the anionic lipid binding is enhanced. Nisin interacts to the anionic liposome of the Gram-positive bacteria, and this interaction is strong because Nisin binds slowly to other liposomes and the fragments of the Nisin helps in identification of the region suitable for binding [68].

Similarly, initially the class 2 bacteriocins also follow the anionic phospholipid membrane interaction.

Membrane Insertion in Food Spoilage Bacteria

As per research study the capability of the Nisin to interact with the lipid layer present in the membrane is enhanced by the presence of a strong antimicrobial activity. It has been stated that a low anti-microbial activity results in a decreased lipid interaction of the bacteriocin nisin and vice versa [69].

Passage through Cell Wall and Binding with Receptors in Food Spoilage Bacteria

The class 1 bacteriocin Lantibiotics are able to form ion conducting pores in the black lipid membranes of the Gram-positive bacteria. As a result of which this results in the interaction with the peptidoglycan precursor lipid 2 in the membrane. The presence of the lipid 2 precursor enhances the ability of the bacteriocin to depolarise the electrical potential of the membrane. These data support the fact that lipid 2 acts as the docking molecule or the nucleus of the membrane which controls all the activities. The Gram-positive cell wall allows the passage of the bacteriocins as enhanced by the Lipid 2 and the polymers present on the cell surface that are lipoic acid and lipoteichoic acid plays a crucial role in the initial interaction of the bacterial cell wall and the bacteriocins. However, receptor plays a role in this whole process and their implications have been explained in many research studies on narrow targets which are specific [70].

By Pore Formation in Food Spoilage Bacteria

The formation of pore has been proposed but particularly for the Lantibiotics-Nisin pore formation it follows a series a specific distinct step. These are first arrangement of Nisin occurs parallelly with the membrane of the Gram-positive bacteria. With the insertion of membrane of the Nisin domains, the cis acid and the trans negative are able to establish pore formation. As the C-Terminus inserts deeply into the membrane the entire Nisin molecule moves across the membrane. Insertion of C-Terminus causes intra-layer attachment with the phospholipids which results in the transmembrane movement of the Nisin across the phospholipid. A PMF (Proton Motive Force) is required and is essential for the creation of the wedge like model by Nisin for pore formation. The creation of wedge like model by Nisin allows bending of C-Terminal part, thereby assisting in membrane insertion. Insertion of multiple Nisin molecules creates a huge local dislocation and a large disruption in the lipid bilayer. As disruption is created in the membrane of the organism it results in the formation of transient pores of lipid [71].

Bacteriocin Affecting DNA Synthesis in Food Spoilage Bacteria

In some of the research of [72] the mechanism of action of megacin C, a bacteriocin of the colin type which is produced by *Bacillus megaterium* has been explained and demonstrated. The insertion of the bacteriocin is generally followed by the inhibition of DNA synthesis in a Gram-positive bacterium along with the inhibition of protein and RNA synthesis. It has been observed that, at very high concentrations of the bacteriocin megacin C, the destruction or inhibition of DNA, as well as protein and RNA synthesis, are quite rapid. However, prior treatment of the bacterium with chloramphenicol and streptomycin decreases the megacin action and limits its activity. As mentioned earlier bacteriocins are protein peptides that bear high anti-bacterial activity which occurs as a chemical component within the cell components of the producing bacteria. It has been established that a single particle is strong enough to kill a strain of sensitive bacteria to the specific receptors sites that are present in the cell wall. Also, according to certain studies and examination about the mode of action of the E. coli bacteriocins that is the Colin group (colicin K & colicin F), an interesting fact about the inhibition of macromolecular synthesis that is protein and nucleic acid synthesis have been revealed. As per studies soon after the adoption of the bacteriocin into the cell membrane of the bacteria, the bacteriocin shuts down the DNA and RNA synthesis without affecting the respiration of the microorganism or the cellular mechanism. Similarly, another research study stated that another colin group of E2 results in complete inhibition of DNA and RNA synthesis in sensitive cells, with a single particle of colicin. As stated by the results of the antimicrobial action of bacteriocin megacin C was established and was concluded that the killing activity of megacin C is just similar to that of a lethal agent. In this experiment, the killing action of megacin C has been established in different concentrations and at different intervals to observe the survival rate. As established from the curve mentioned the viable count of the bacterial cells subsequently decreased and the rate of killing proportionately increased with increasing concentrations of the bacteriocin, megacin C which proved the fact that adsorption by a single particle is lethal enough to kill or cause destruction [72]. Similarly, the effect of bacteriocin on the macromolecular synthesis that is protein and nucleic acid synthesis was also established by [72], who stated that previously the bacterial strains of *B. megaterium* were treated with different concentrations of megacin C and the survival rate was observed. Thus, all the samples were at intervals and the number of nucleic acids and protein were determined. Inhibition of DNA was observed as the concentration of megacin applied was higher. Similarly, the protein and nucleic acid synthesis were also inhibited when the concentrations were increased; however, the effect was indirect, which was consistent with the DNA dependant RNA synthesis.

Other Factors Affecting Bacteriocin Production

LAB is typically used as a starter culture in the fermentation of foods and beverages as it helps in yielding a good flavor and aroma to the food and therefore assisting in preventing its deterioration. Also, the LAB helps in breaking up macronutrients like carbohydrates, protein, and fat and thus helps in the absorption of nutrients which keeps good health [73]. As a result, LAB is considered to be safe for human consumption which has been consumed through fermented foods for several years. The LAB is mainly applied in the food industry for food preservation. In most, industrialized nations around 30% of the total population is getting affected by bacterial pathogens or fungi which cause food spoilage and make the food unfit for human consumption. As a result of which chemical preservatives are used for food preservation having severe side effects on health in the long term [74]. Therefore, the food bio-preservation techniques are using bacteriocin as an alternative to the harsh chemical preservatives to protect the food against bacterial and fungal spoilage. Also, it does not affect human health adversely in any way. A research study discusses several different factors affecting bacteriocin production by *Lactobacillus* bacteria. The biosynthesis of bacteriocin depends upon various conditions of the culture medium, such as pH, growth kinetics of microorganisms, and temperature. But a challenge faced is the economic cost which is needed for the large-scale production of bacteriocin in the food industry for food preservation; therefore it should be taken into consideration by finding out the optimal conditions needed for bacteriocin production. According to, there are many culture conditions like pH, temperature, the composition of the medium, growth kinetics of the microorganisms which affect the production of bacteriocins. In many cases, the maximum production of bacteriocin occurs at different conditions of optimum condition for the growth of microorganisms.

Conclusion

The bacteriocins are the group of antimicrobial peptides produced by various types of bacterial cells in various types of microecological systems. The bacteriocins being produced by LAB are the groups of natural bio preservatives help in preserving the food against various types of resistant Gram-positive as well as Gram-negative bacterial cells. The advent of the bacteriocin would reduce the use of antibiotics and enhance their activity as natural food preservatives. The use of specific bacteriocins does not possess the ability of eliminating all possible types of pathogenic bacterial cells thereby mixing varieties of antimicrobial peptides like bacteriocin may prevent the growth of dreadful pathogenic cells. In coming years the increase in multidrug resistant pathogens may result in the development of serious consequences thus development of new generation of antimicrobials is of utmost importance. Use of metagenomic analysis may open a new avenue to find out new bacteriocin producing bacterial groups. Application of system biology and other biotechnological approaches including the recombinant

DNA technologies may result in the development of new types of bacteriocin which might have better impact upon the resistant cells.

References

[1] Zhang J. C., Sun L., Nie Q. H., (2010). Botulism, where are we now? *Clin. Toxicol.* (Phila). 48, 867–879.

[2] Eijlander R. T., Abee T., Kuipers O. P., (2011). Bacterial spores in food: how phenotypic variability complicates prediction of spore properties and bacterial behavior, *Curr. Opin. Biotechnol.* 22, 180–186.

[3] Müller-Herbst S., Wüstner S., Mühlig A., Eder D., Fuchs T. M., Held C., Ehrenreich A., Scherer S., (2014). Identification of genes essential for anaerobic growth of Listeria monocytogenes, *Microbiol.* 160, 752–765.

[4] Elsner H. A., Sobottka I., Mack D., Claussen M., Laufs R., Wirth R., (2000). Virulence factors of Enterococcus faecalis and Enterococcus faecium blood culture isolates, *Eur. J. Clin. Microbiol. Infect. Dis.* 19, 39–42.

[5] Gautam, N., Sharma, N. (2009). Bacteriocin: safest approach to preserve food products. *Indian Journal of Microbiology,* 49(3), 204–211. https://doi.org/10.1007/s12088-009-0048-3.

[6] Todorov, S. D., Leblanc, J. G., Franco, B. D. G. M., (2012). Evaluation of the probiotic potential and effect of encapsulation on survival for Lactobacillus plantarum ST16Pa isolated from papaya, *World J. Microbiol. Biotechnol.* 28, 973–984.

[7] Cleveland, J., Montville, T. J., Nes, I. F., Chikindas, M. L. (2001). Bacteriocins: safe, natural antimicrobials for food preservation. *Int J Food Microbiol,* 71, 1-20.

[8] Riley, M. A., Wertz, J. E. (2002). Bacteriocins: evolution, ecology, and applications. *Annu Rev Microbiol,* 56, 117-137.

[9] Cotter, P. D., Hill, C., Ross, R. P. (2005). Bacteriocins: Developing innate immunity for food. *Nat Rev Microbiol,* 3, 777-788.

[10] Diep, D. B., Nes, I. F. (2002). Ribosomally synthesized antibacterial peptides in Gram-positive bacteria. *Curr Drug Targets,* 3, 107-122.

[11] Rodriguez E., Martinez M. I., Horn N., Dodd H. M., (2003). Heterologous production of bacteriocins by Lactic Acid Bacteria. *International Journal of Food Microbiology,* 80, 101-116.

[12] Rodriguez E. G. B., Gaya P., Nanez M., Medina M., (2000). Diversity of bacteriocins produced by Lactic Acid Bacteria isolated from raw milk. *International Dairy Journal,* 10, 7-15.

[13] Moll G. N., Konings W. N., Driessen, A. J. M., (1999). Bacteriocins: mechanism of membrane insertion and pore formation. *Antonie van Leeuwenhoek Journal,* 3, 185-195.

[14] Silva Célia C. G., Silva Sofia P. M., Ribeiro Susana C., (2018). Application of bacteriocins and protective cultures in dairy food preservation, *Frontiers in Microbiology,* 9, 594.

[15] Todorov, S. D., Dicks L. M. T., (2005). Lactobacillus plantarum isolated from molasses produces bacteriocins active against Gram negative bacteria. *Enzyme and Microbial Technology Journal,* 36, 318-326.

[16] Todorov S. D., Dicks L. M. T., (2004). Influence of growth conditions on the production of a bacteriocin by Lactococcus lactis subp. lactis ST 34BR, a strain isolated from barley beer. *Journal of Basic Microbiology,* 44, 305-316.

[17] Todorov S. D. D., Dicks L. M. T., (2005). Effect on Growth medium on bacteriocin production by Lactobacillus plantarum ST194BZ, a strain isolated from boza. *Journal of Food Technology and Biotechnology,* 43, 165-173.

[18] Aguilar, A. (1991). Biotechnology of lactic acid bacteria: An European perspective. *Food Biotechnol.* 5, 323-330.

[19] Stiles, M. E. (1996). Biopreservation by lactic acid bacteria. *Anton van Leeuw* 70: 331-345.

[20] Marrug, J. D. (1995). Bacteriocins, their role in developing natural products. *Food Biotechnol.* 5: 305-312.

[21] Holzapfel, W. H., R. Geisen, and U. Schillinger (1995). Biological preservation of foods with reference to protective cultures, bacteriocins and food-grade enzymes. *Int. J. Food Microbiol.* 24: 343-362.

[22] Holtzel, A., M. G. Ganzle, G. J. Nicholson, W. P. Hammes, and G. Jung (2000). The first low-molecular-weight antibiotic from lactic acid bacteria: Reutericyclin, a new tetramic acid Angewandte. *Chem. Int. Edi.* 39: 2766-2768.

[23] Magnusson, J. and J. Schnürer (2001). Lactobacillus coryniformis subsp Coryniformis strain Si3 produces a broad-spectrum proteinaceous antifungal compound. *Appl. Environ. Microbiol.* 67: 1-5.

[24] Daeschel, M. A. (1993). Applications and interactions of bacteriocins from lactic acid bacteria in foods and beverages. pp. 63-91. In: D. G. Hoover and L. R. Steenson (eds.). *Bacteriocins of Lactic Acid Bacteria.* Academic Press Inc.

[25] Gálvez, A., Abriouel, H., López, R. L., Omar, N. B., (2007). Bacteriocin-based strategies for food biopreservation. *Int. J. Food Microbiol.* 120: 51-70.

[26] Dodd, H. M. and M. J. Gasson (1994) *Bacteriocins of Lactic Acid Bacteria*. pp. 211-255. In: M. J. Gasson and W. M. de Vos (eds.).

[27] Genetics and biotechnology of lactic acid bacteria. Blackie Academic and Professional, London, UK. Piard, J. C. and M. J. Desmazeaud (1992) Inhibiting factors produced by lactic acid bacteria 2 Antibacterial substances and bacteriocins. *Lait.* 72: 113-142.

[28] Ng, Z. N., Zarin, M. A., Lee, C. K., Tan, J. S. (2020). Application of bacteriocins in food preservation and infectious disease treatment for humans and livestock: a review. *RSC Advances,* 10, 38937–38964.

[29] Zacharof, M. P., Lovitt, R. W. (2012). Bacteriocins produced by lactic acid bacteria: A review article. *APCBEE Procedia,* 2, 50-56.

[30] Perez, R. H., Zendo, T., Sonomoto, K. (2014). Novel bacteriocins from lactic acid bacteria (LAB): various structures and applications. *Microbial Cell Factories,* 13, (Suppl1): S3.

[31] Yao, W., Yuxuan, Q., Qing, X., Ying, Z., Jinrong, H., Pinglan, L. (2018). Purification and characterization of plantaricin LPL-1, a novel class IIa bacteriocin produced by Lactobacillus plantarum LPL-1 isolated from fermented fish, *Frontiers in Microbiology,* 9, 2276.

[32] Elayaraja, S., Annamalai, N., Mayavu, P., Balasubramanian, T., (2014). Production, purification and characterization of bacteriocin from Lactobacillus murinus AU06 and its broad antibacterial spectrum, *Asian Pacific Journal of Tropical Biomedicine,* 4(1), S305-S311.

[33] Ishibashi, N., Himeno, K., Fujita, K., (2014). Purification and characterization of multiple bacteriocins and an inducing peptide produced by Enterococcus faecium NKR-5-3 from Thai fermented fish, *Bioscience, Biotechnology, and Biochemistry,* 76(5), 947–953.

[34] Szabóová, R., Lauková, A., Simonová, M. P., Strompfová, V., Chrastinová, L., (2012). Bacteriocin-producing enterococci from rabbit meat, *Malaysian Journal of Microbiology,* 8(4) 211–218.

[35] Biscola, V., Todorov, S. D., Capuano, V. S. C., Abriouel, H., Gálvez, A., Franco, B. D. G. M., (2013). Isolation and characterization of a nisin-like bacteriocin produced by a Lactococcus lactis strain isolated from charqui, a Brazilian fermented, salted and dried meat product, *Meat Science,* 93 (3) 607–613.

[36] Zhang, H., Liu, L., Hao, Y., (2013). Isolation and partial characterization of a bacteriocin produced by Lactobacillus plantarum BM-1 isolated from a traditionally fermented Chinese meat product, *Microbiology and Immunology,* 57(11), 746–755.

[37] Hu, Y., Liu, X., Shan, C., (2017) .Novel bacteriocin produced by Lactobacillus alimentarius FM-MM 4 from a traditional Chinese fermented meat Nanx Wudl: purification, identification and antimicrobial characteristics, *Food Control,* 77, 290–297.

[38] Lv, X., Ma, H., Sun, M., (2018). A novel bacteriocin DY4-2 produced by Lactobacillus plantarum from cutlassfish and its application as bio-preservative for the control of Pseudomonas fluorescens in fresh turbot (Scophthalmus maximus) fillets, *Food Control,* 89, 22–31.

[39] Zhang, J., Yang, Y., Yang, H., Bu, Y., Yi, H., Zhang, L., Han, X., Ai, L. (2018). Purification and partial characterization of bacteriocin Lac-B23, a novel bacteriocin production by Lactobacillus plantarum J23, isolated from Chinese traditional fermented milk. *Frontiers in Microbiology,* 9, 2165. https://doi.org/10.3389/fmicb.2018.02165.

[40] Lahiri, D., Chakraborti, S., Jasu, A., Nag, M., Dutta, B., Dash, S., Ray, R. R., (2020). Production and purification of bacteriocin from Leuconostoc lactis SM 2 strain, *Biocatalysis and Agricultural Biotechnology*, Volume 30, 101845.

[41] Klaenhammer, T. R. (1993). Genetics of bacteriocins produced by lactic acid bacteria. *FEMS Microbiol Rev,* 12, 39-85. Diep, D. B., Nes, I. F. (2002). Ribo-

somally synthesized antibacterial peptides in Gram-positive bacteria. *Curr Drug Targets,* 3, 107-122.

[42] Cotter, P. D., Hill, C., Ross, R. P. (2005). Bacteriocins: Developing innate immunity for food. Nat Rev Microbiol, 3, 777-788.

[43] Rodríguez, J. M., Martínez, M. I., Horn, N., Dodd, H. M. (2003). Heterologous production of bacteriocins by lactic acid bacteria. *Int J Food Microbiol.* 80(2), 101-116. doi:10.1016/s0168-1605(02)00153-8.

[44] Zacharof, M. P., Lovitt, R. W. (2012). Bacteriocins produced by lactic acid bacteria: A review article. *APCBEE Procedia,* 2, 50-56.

[45] Cleveland, J., Montville, T. J., Nes, I. F., Chikindas, M. L. (2001). Bacteriocins: safe, natural antimicrobials for food preservation. *Int J Food Microbiol.* 71(1),1-20. doi:10.1016/s0168-1605(01)00560-8.

[46] Zacharof, M. P., Lovitt, R. W. (2012). Bacteriocins produced by lactic acid bacteria: A review article. *APCBEE Procedia,* 2, 50-56.

[47] Cotter, P. D., Hill, C., Ross, R. P. (2005). Bacteriocins: developing innate immunity for food. *Nat Rev Microbiol,* 3, 777-788.

[48] Daw, M. A., Falkiner, F. R. (1996). Bacteriocins: nature, function and structure. *Micron,* 27(6), 467–479. doi:10.1016/s0968-4328(96)00028-5.

[49] Maqueda, M., Sanchez-Hidalgo, M., Fernandez, M., Montalban-Lopez, M., Valdivia, E., Martinez-Bueno, M. (2008). Genetic features of circular bacteriocins produced by Gram-positive bacteria. *FEMS Microbiol Rev,* 32, 2-22.

[50] Klaenhammer T. R. 1993. Genetics of bacteriocins produced by lactic acid bacteria. *FEMS Microbiol. Rev.* 12:39-86.

[51] Fath, M. J., Zhang, L. H., Rush, J., Koiter, R., (1994). Purification and characterization of colicin V from Eschericia coli culture supernatants. *Biochemistry* 33:6911-6917.

[52] Havarstein, L. S., Coomaraswamy, G., Morrison, D. A., (1995). An unidentified heptadecapeptide pheromone induces competence for genetic transformation in Streptococcus pneumoniae. Proc. Natl. Acad. Sci. USA 92: I 1140-11144.

[53] Nissen-Meyer, J., Holo, H., Htvarstein, L. S., Sletten, K., Nes, I. F., (1992). A novel lactococcal bacteriocin whose activity depends on the complementary action of two peptides. *J. Bacteriol.* 174: 5686-5692.

[54] Abee, T., (1995). Pore-forming bacteriocins of Gram-positive bacteria and self-protection mechanisms of producer organisms. *FEMS Microbiol. Lett.* 129:1-10.

[55] Leer, R. L., van der Vossen, J. M. B. M., van Giezen, M., van Noort, J. M., Pouwels, P. H., (1995). Genetic analysis of acidocin B, a novel bacteriocin produced by Lactobacillus acidophilus. *Microbiol.* 141:1629-1635.

[56] Riley, M. A., Wertz, J. E. (2002). Bacteriocins: evolution, ecology, and application. *Annu Rev Microbiol.* 56:117-137. doi:10.1146/annurev.micro.56.012302.161024.

[57] Klaenhammer, T. R. (1993). Genetics of bacteriocins produced by lactic acid bacteria. *FEMS Microbiol Rev.* 12(1-3):39-85. doi:10.1111/j.15746976.1993.tb000 12.x.

[58] Stevens, K. A., Sheldon, B. W., Klapes N. A., Klaenhammer T. R. (1991). Nisin treatment for inactivation of Salmonella species and other gram-negative bacteria. *Appl Environ Microbiol.* 57(12):3613-3615. doi:10.1128/AEM.57.12.36133615.1991.

[59] Breukink, E., Wiedemann, I., van Kraaij, C., Kuipers, O. P., Sahl, H. G., de Kruijff, B. (1999). Use of the cell wall precursor lipid II by a pore-forming peptide antibiotic. *Science.* 286(5448):2361-2364. doi:10.1126/science.286.5448.2361.

[60] Fimland, G., Johnsen, L., Dalhus, B., Nissen-Meyer, J. (2005). Pediocin-like antimicrobial peptides (class IIa bacteriocins) and their immunity proteins: biosynthesis, structure, and mode of action. *J Pept Sci.* 11(11):688-696. doi:10.1002/psc.699.

[61] Oppegård, C., Kjos, M., Veening, J. W., Nissen-Meyer, J., Kristensen, T., (2016). A putative amino acid transporter determines sensitivity to the two-peptide bacteriocin plantaricin JK. *Microbiologyopen.* 5(4):700-708. doi:10.1002/mbo3.363.

[62] Garneau, S., Martin, N. I., Vederas, J. C. (2002).Two-peptide bacteriocins produced by lactic acid bacteria. Biochimie. 84(5-6):577-592. doi:10.1016/s0300-9084(02)01414-1.

[63] Oppegård, C., Kjos, M., Veening, J. W., Nissen-Meyer, J., Kristensen T. (2016). A putative amino acid transporter determines sensitivity to the two-peptide bacteriocin plantaricin JK. *Microbiologyopen.* 5(4):700-708. doi:10.1002/mbo3.363.

[64] Gabrielsen, C., Brede, D. A., Hernández, P. E., Nes, I. F., Diep, D. B. (2012). The maltose ABC transporter in Lactococcus lactis facilitates high-level sensitivity to the circular bacteriocin garvicin ML. *Antimicrobial Agents and Chemotherapy,* 2908-2915. doi:10.1128/AAC.00314-12.

[65] Butlin, J. D., Cox, G. B., Gibson F. (1971). Oxidative phosphorylation in Escherichia coli K12. Mutations affecting magnesium ion- or calcium ion-stimulated adenosine triphosphatase. *Biochemical Journal,* 124(1), 75-81. doi:10.1042/bj1240075.

[66] Maloney, P. C., Kashket, E. R., Wilson, T. H. (1974). A protonmotive force drives ATP synthesis in bacteria. *PNAS,* 71, 3896-3900. doi: 10.1073/pnas.71.10.3896.

[67] Bruno, M. E. C., Montville, T. J., (1993). common mechanistic action of bacteriocins from lactic acid bacteria. *Applied and Environmental Microbiology,* 59(9), 3003–3010.

[68] Wayah, S. B., Philip, K., (2018). Pentocin MQ1: A novel, broad-spectrum, pore-forming bacteriocin from Lactobacillus pentosus CS2 with quorum sensing regulatory mechanism and biopreservative potential. *Frontiers In Microbiology,* 9. doi:10.3389/fmicb.2018.00564.

[69] Driessen, A. J. M., van den Hooven, H. W., Kuiper, W., de Camp, M. V., Sahl, H. G., Konings, R. N. H., Konings, W. N. (1995). Mechanistic studies of lantibiotic-induced permeabilization of phospholipid vesicles. *Biochemistry,* 34, 1606–1614.

[70] Demel, R. A., Peelen, T., Siezen, R. J., Kruijff, B. D., Kuipers, O. P., (1996). Nisin Z, Mutant Nisin Z and Lacticin 481 Interactions with anionic lipids correlate with antimicrobial activity. *European Journal of Biochemistry,* 267–274.

[71] Jack, R. W., Tagg, J. R., Ray, B. (1995). Bacteriocins of gram-positive bacteria. *Microbiology Reviews,* 59(2), 171–200.

[72] Eefjan, B., van Kraaij, C., van Dalen, A., Demel, R. A., Siezen, R. J., de Kruijff, B., Kuipers, O. P., (1998). The Orientation of Nisin in Membranes. *Biochemistry,* 8153–8162.

[73] Holland I. B. (1965). A Bacteriocin specifically affecting DNA Synthesis in Bacillus megaterium. *Journal of Molecular Biology,* 12, 429-438.

[74] Gemechu T. (2015). Review on lactic acid bacteria function in milk fermentation and preservation. *Africal Journal of Food Science,* 9(4), 170-175.

[75] Bharti, V., Mehta, A., Singh, S., Jain, N., Ahirwal, L., Mehta, S. (2015). Bacteriocin: A novel approach for preservation of food. *International Journal of Pharmacy and Pharmaceutical Sciences,* 7(9), 1-10.

Chapter 9

Positive Impact of a Novel Organic Fertilizer on the Plant Health and the Nutritional Value of Okra (*Abelmoschus Esculentus* L.) in Comparison to Chemical Fertilizer

Susmita Mukherjee and Pratik Talukder*
Department of Biotechnology,
University of Engineering and Management,
Kolkata, West Bengal, India

Abstract

Okra (*Abelmoschus esculentus* L.) is an important nutrition rich vegetable crop of malvaceae family. The modern system of farming by using more and more chemical fertilizers is becoming unsustainable which is quite evident from the declining plant health, lesser nutritional values of the crops, reduced crop productivity and above all damage to environment. The necessity of having an alternative agriculture practice by using organic fertilizers is realized now. The objective of this study is to establish the use of organic fertilizers as a viable alternative to chemical fertilizers. Hence a comparative study between the effect of organic fertilizers and chemical fertilizers in terms of plant health, growth, nutritional value and secondary metabolite production in okra was carried out. In this study it was observed that the okra plants which were cultivated by using only organic fertilizer showed higher amount of carbohydrate and protein production compared to the plants grown in chemical fertilizer administered fields. It also showed that the plants grown in organic fertilizer treated fields are less stressed. Hence it can be

* Corresponding Author's Email: pratik.talukder@uem.edu.in.

In: Proceedings of BIOSPECTRUM
Editors: Sanket J. Joshi, Susmita Mukherjee and Moupriya Nag
ISBN: 978-1-68507-985-7

propounded from the above observations that organic fertilizers are playing beneficial role in alleviating stress in okra plants as the level of several stress markers are low compared to that in the plants grown in field administered with chemical fertilizers. The outcome of this study clearly indicates the positive role of organic fertilizers in enhancing the nutritional values in okra and also emphasizes about the benefits of consuming vegetables grown by using organic fertilizers.

Keywords: okra, *Abelmoschus esculentus* L., organic fertilizers, chemical fertilizers, secondary metabolites, antioxidant

Introduction

Food habit in human has evolved over the course of human evolution and human civilization. One of the main ingredients of our diet is vegetables. In India okra (*Abelmoschus esculentus* L.) is an important part of our daily diet as it is a rich source of macro and micro nutrient (Lyngdoh et al. 2013). This erect annual herb belongs to the malvaceae family. Not only in India but also it is consumed worldwide. Several studies showed that apart from being a rich source of protein, carbohydrate and microelements it also serves as a potent producer of secondary metabolites such as polyphenols and flavonoids (Yang et al. 2012). Being a fast growing developing nation with such a huge population it is of utmost importance to increase the production of vegetables in India but at the same time we should also focus enhancing the nutritional value of the vegetables. Environmental alterations outside the optimal range results in the stress related loss of crop yield. Hence the entire focus is now shifted towards production enhancement. By green revolution higher productivity of crops and vegetables was achieved but unfortunately in the quest of increasing yield the quality and nutritional aspects of the vegetables are greatly compromised because of indiscriminate use of chemical fertilizers, pesticides (Mahajan et al. 2008; Hijbeek et al. 2018). Apart from this very many indigenous varieties are also lost due to this (Nelson et al. 2019). Now a days food safety has been a primary concern of scientists worldwide. In a highly populated country like India the need of the hour should not be only aimed at enhancement of food production but also in maintaining and if possible enhancing the nutritional values of agricultural products and maintenance of plant health (Yamaguchi & Hwang, 2015). Rapid industrialization and urbanization has taken a toll on the agricultural lands and soil health. Total arable land is decreasing day by day. Farmers are now growing vegetables in small hamlets nearer to the suburban areas and using higher amount of chemical fertilizers to maximize the production. These chemical fertilizers are not only affecting the health of the vegetable plants but also get accumulated in the soil and water thus causing huge negative impact on our environment (Sharma & Singhvi, 2017). Unscrupulous use of pesticides has become a practice. Although this contributes to higher yield to some extent but contributes to nutrient imbalance, production of reactive oxygen and nitrogen species (ROS and RNS) and eventually brings about deterioration of crop and soil quality (Mal et al. 2013). Now it is the high time to curb the wanton use of chemical fertilizers. The main dilemma before us is how to reduce the use of chemical fertilizers and replace it with organic fertilizers without compromising the yield and nutritional value. One the approaches which can be adopted to enhance nutritional values of the products without harming the plant health and also minimizing the ill effect on the people who are consuming these vegetables is to switch to the use of organic fertilizers. In a recent study on the effect of organic fertilizers on yield of okra it was reported that in terms of fruit size and fruit number a significant increase was observed under the influence of organic fertilizers but the impact on nutritional values and parameters of plant health was not reported [9]. This field study on okra reports the positive impact of organic fertilizers on nutritional values of okra because apart from increasing yield the most important aspect is nutritional value of it as we consume vegetables to have nutrition from them. Phenolics in fruits and vegetables represent the major class of antioxidant supplements in our diet (Liang & Kitts, 2016). Plants produce secondary metabolites mainly via phenylpropanoid pathway. The first step of this pathway is accomplished by Phenylalanine ammonia Lyase (PAL) (Goujon et al. 2003) Polyphenolic compounds are one of the major groups among the various secondary metabolites which can quench the free radicals produced due to various biotic and abiotic stresses (Dixon and Paiva, 1995). There are several defense mechanisms evolved in plants to withstand stress induced toxicity including the production of natural antioxidants such as polyphenols (Mittler, 2002). Several studies documented that okra is a good source of natural antioxidants such as polyphenols (Ahiakpa et al. 2013). In this present field study a detailed analysis on how the natural antioxidant secondary metabolites play important role in mitigating stress and thus help in maintaining plant health was carried out. One of the main deterrents is cost; high cost of organic fertilizers compared to that of chemical fertilizers deters the farmers from using it. In this study we have formulated a noble organic fertilizer which can be marketed in affordable low price. Although it seems to be a tedious job we have to keep trying to make the farmers understand the beneficial aspects of using organic fertilizers. The main aim should be to increase the use of organic fertilizers keeping in mind that the plant health and hence the nutritional value of the plant should not be compromised. In this present study the analysis was mainly focused on plant health, stress management and nutritional aspect of okra.

Methods

Plant Materials

Seeds of Okra (*Abelmoschus esculentus*) [Variety- BCO- 1] were collected from Bidhan Chandra Krishi Viswavidyalaya, West Bengal, India. Plants were grown from these seeds in three fields located in a same area. The type of fertilizers used in those fields is as follows:

- In Field – 1 neither organic fertilizers nor chemical fertilizers were used (Control group – F1)
- In Field – 2 only chemical fertilizer was used (Chemical fertilizer group – F2).
- In Field – 3 only organic fertilizer was used (Organic fertilizer group – F3).

Table 1. Concentration of different macro and micro elements present in soil where the plants were grown

Parameter	Method of Analysis	Unit	Concentration
Nitrogen	IS: 14684	mg/g	0.65
Potassium	IS: 9497	mg/g	1.8
Phosphorous	IS: 5305	mg/g	0.39
Iron	APHA: 3500 FeB	mg/g	0.08
Sulphur	IS: 2720 (Part XXVII)-1977	%	0.04
Boron	4500BC	mg/kg	0.21
Zinc	APHA3111B	mg/g	0.03

The important macro and micro elements were tested and on the basis of the report the dosage and proportion of each component in both organic and chemical fertilizer was finalized. The result of soil health report is given in Table 1.

Composition of Organic Fertilizer Used in This Study

Compost, mustard cake, neem cake, bone dust was mixed in 5:1:1:2 ratios and the mixture were administered in the field.

Composition of Chemical Fertilizer Used in This Study

NPK fertilizer (10:10:10), magnesium sulphate, ferrous sulphate, zinc sulphate mixture was used as chemical fertilizers in 5:1:2:2:2 ratios.

Three different fields were prepared. In average thirty plants were maintained in each field.

Alteration of Seedling Height

The reduction in shoot and root length can be caused due to stress induction by the harmful chemicals present in higher than permissible concentration in the chemical fertilizers which are used indiscriminately in the agriculture fields. Shoot length of two month old plants were calculated and expressed as mean ± standard error of mean (SEM).

Estimation of Total Chlorophyll and Carotenoids

Amount of Chlorophyll was measured according to the method of Lichtenthaler, 1987 with minor modifications. 500 mg plant tissue was crushed in 100% acetone solution. The homogenate was centrifuged at 12000 rpm for 15 minutes. OD was recorded and with the help of the following formula content of chlorophyll and carotenoid was estimated:

Total Chlorophyll= OD at 663 nm X 7.15 + OD at 646 nm X 18.71

Total Carotenoids (Including both Carotene and Xanthophyll)

= {(OD at 470 nm X 1000) – (Chl a content X 1.90) – (Chl b content X 63.14)}/ 214

Estimation of Proline

Total proline content was calculated according to the method described by Bates et al. (1973) with minor modifications. 500 mg of plant was homogenized in 3 ml of 5% (w/v) tri-chloro acetic acid. The homogenate was then centrifuged for 5 min at 10,000 rpm. 2 ml of supernatant (2 ml) was collected and mixed with 2 ml of acid ninhydrin, ortho-phosphoric acid and glacial acetic acid solution. These constituents were mixed thoroughly by vortex and after that boiled at 90°c for 45 minutes. The resulting product was rapidly cooled on ice to stop the reaction. 4 ml of toluene was added to 1 ml of this product and shaken vigorously for 15 – 20 seconds to form a homogenous solution. The solution was kept at room temperature for 15 minutes for phase separation. The chromophore containing phase was taken and absorbance was recorded at 520 nm. Total proline content (mg/g fresh weight of tissue FW) was calculated from that standard curve of proline.

Estimation of Extent of Lipid Peroxidation

Procedure of Heath & Packer, (1968) with little modifications was adapted. 500 mg tissue was homogenized in 4 ml of Tri-chloro acetic acid and Thiobarbituric Acid mixture. The solution was centrifuged at 15000 rpm (10 minutes). It was then boiled for 40 minutes and kept for 1 hour in room temperature. After cooling at room temperature OD was taken at 532 nm.

Preparation of Plant Extract

Brolis et al. (1998) was followed with little modifications. 1 gm tissue was crushed using ethanol (50%) and centrifuged for 10 minutes at 10000 rpm. The supernatant was further sonicated in bath sonicator for 20 minutes and was stored in - 20°C temperature for future use.

Estimation of Phenolics

Determination of Total Polyphenol Content

Total polyphenol content was estimated according to Singleton et al. (1999) with few modifications (Talukder et al. 2016). 100 µl plant extract, 300 µl Folin- ciocalteu reagent and 600 µl 10% (W/V) sodium carbonate solution was mixed and incubated in dark at room temperature for 45 minutes. OD was taken at 760 nm. Total polyphenol content was calculated by comparing with the gallic acid calibration curve.

Determination of Total Flavonoid Content

Total flavonoid content was determined by method described by Lin & Tang, (2007) with few modifications (Talukder et al. 2016). 500 µl plant extract, 100 µl of 10% (w/v) aluminium trichloride hexahydrate, 100 µl 1M potassium acetate and 2.3 ml deionized water was mixed thoroughly. Incubation was done for 40 minutes at room temperature. OD was recorded in at 415 nm. With the help of a rutin calibration curve total flavonoid content was measured.

DPPH Radical Scavenging Assay

Brand-Williams et al. (1995) was followed with few modifications. 0.7 ml plant extract was mixed with 2.3 ml of reagent solution and incubation was done in dark condition for two hours.

The extent of radical scavenging was measured as:

Percentage activity = [(OD of control- OD of sample)/OD of control] X 100%

Phenylalanine Ammonia Lyase (PAL) Activity Assay

Activity of PAL was measured according to the method of Dos Santos et al. (2004) with few modifications (Kova´cˇik et al. 2007). To prepare 100 ml 50 mM L-phenylalanine 0.825g of L-phenylalanine was weighed and taken in a conical flask. To it, 80 ml distilled water was added and mixed gently. The volume was finally made up to 100 ml using distilled water and it was stored at - 20°C until further use. Plant tissue (1 gm) was homogenized at 4°C in 0.1 M sodium borate buffer (pH 8.8). Centrifugation was carried out at 15000 rpm for 20 minutes. To 500 µl of this crude enzyme 500 µl 100 mM phenylalanine solution was added as substrate to initiate the enzyme substrate reaction. Incubation time for this reaction to occur was 60 minutes. Finally reaction was terminated by adding 30 µl of 5N HCl. OD was taken 275 nm which corresponds with the absorption maxima of cinnamic acid; the product of this PAL catalysed reaction.

Determination of Total Protein Content

Protein estimation was done according to the method given by Okutucu (2017). 500 µl of plant extract was mixed with 500 µl distilled water and 1 ml of alkaline copper sulphate solution and mixed thoroughly. After 45 minute incubation at room temperature 250 µl of Folin's reagent was added and again incubated for another 30 minutes. Finally Absorbance was recorded using spectrophotometer at 660nm. Concentration of protein was estimated as BSA equivalent by using BSA Standard curve.

Determination of Total Carbohydrate Content

Carbohydrate assay was done according to the method of Hedge & Hofreiter (1962) with few modifications. 4ml of freshly prepared anthrone reagent and 400µl of ethanolic plant extract was mixed and the reaction mix was incubated at 65°C for 10 minutes. The samples were cooled down and subjected to spectroscopic analysis at 620nm. Concentration of carbohydrate was estimated as glucose equivalent by using D Glucose Standard curve.

Statistical Analysis

Data analysis was done by an analysis of variance (ANOVA) and Student's t test. Data are represented as mean ± standard error and level of significance is given as $p < 0.05$ (*); $p < 0.01$ (**); $p < 0.001$(***).

Result and Discussion

A comparative analysis was carried out between the impact of using chemical and organic fertilizers on plant health and nutritional aspect.

Physical Parameter

Alteration of Seedling Height

Average height of the plants was estimated to get an idea of the growth of the plants under the influence of fertilizers (Figure 1). It was observed that average plant height remained unaltered in the plants grown in the field administered with chemical fertilizers (F2) compared to that of the plants grown in filed without any fertilizers (F1). However a significant increase in plant height was recorded in the plants which were grown in the field supplemented with organic fertilizers (F3). A positive correlation between use of manure and increase in shoot height of okra was reported by Tiamyu et al. (2012); a similar positive impact of our novel organic fertilizer on shoot height of okra was observed in the present study. The height of plant is a morphological marker which is a direct manifestation of plant health and growth.

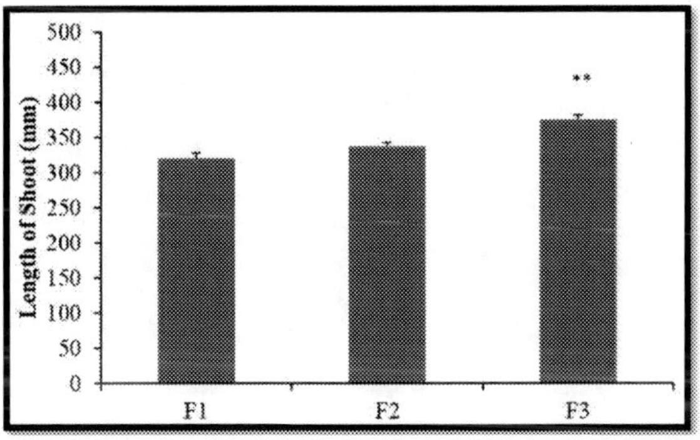

Figure 1. Changes in shoot length (mm) of Okra (*Abelmoschusesculentus*).
Control group- F1, Chemical fertilizer group- F2, Organic bio fertilizer group-F3.
Data are represented as mean ⊥ standard error.
Level of significance: $p < 0.05$ (*); $p < 0.01$ (**); $p < 0.001$(***).

Biochemical Parameters

Some important stress markers such as chlorophyll and proline content, extent of lipid peroxidation were studied to assess the magnitude of stress on Okra.

Total Chlorophyll and Carotenoid Content

These are assayed as plant stress marker. Wanton use of chemical fertilizers can have negative impact on several biochemical pathways and even it can lead to reduction in the rate of photosynthesis and reduction of chlorophyll production. Hence in this present study, Chlorophyll and carotenoid content was estimated to check the impact of chemical and organic fertilizers on okra. Both cholorophyll (Figure 2) and carotenoid (Figure 3) content showed the same pattern of accumulation. In the control F1 plants average chlorophyll content was recorded as 33.23 mg/l which decreased a bit and became 23.05 mg/l in F2 plants. This reduction in chlorophyll content is the indicator of negative impact of chemical fertilizers on plant health. Both chlorophyll and carotenoid content increased in F3 plants (organic fertilizer administered) and recorded as 48.45 mg/l and 32.37 mg/l respectively. This result signifies the positive impact of organic fertilizers on plant health.

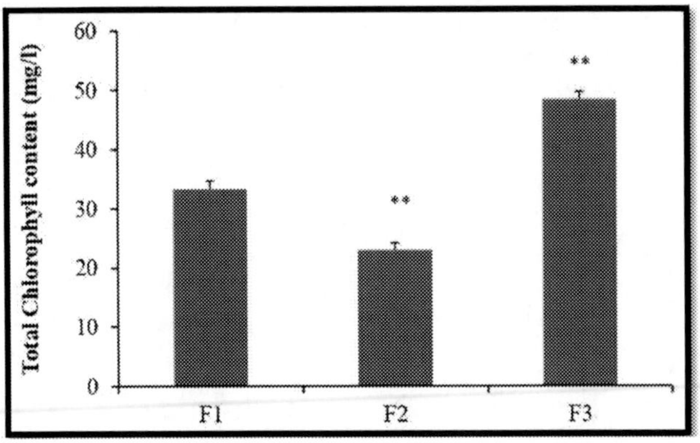

Figure 2. Changes in total chlorophyll content in Okra (*Abelmoschusesculentus*). Control group- F1, Chemical fertilizer group- F2, Organic bio fertilizer group-F3. Data are represented as mean ± standard error.
Level of significance: $p < 0.05$ (*); $p < 0.01$ (**); $p < 0.001$(***).

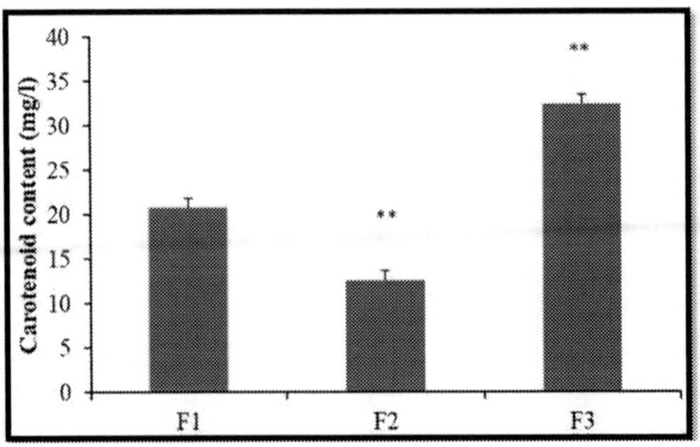

Figure 3. Changes in total carotenoid content in Okra (*Abelmoschusesculentus*).

Total Proline Content

Plants accumulate proline in the cytoplasm under stressed condition. Total proline content in F1 plants was calculated as 56.76 μmol/g FW, it jumped to 96.5 μmol/g FW in F2 (chemical fertilizer administered) plants. It clearly indicates that chemical fertilizers induced stress in the plants. The proline content declined significantly and became 35.35 μmol/g FW in F3 plants (organic fertilizer administered) which is even lesser than the proline content in control plants (Figure 4). Hence it can be propounded from the above observations that organic fertilizers are having beneficial impact on plant health hence the stress markers are accumulated in quite low amount in plants grown in fields in which only organic fertilizers are used. According to a study by Agarwal & Pandey (2004), proline content increased with induction of abiotic stress in *Cassia angustifolia*. The findings of our study also showed a significantly high proline accumulation in plants grown in chemical fertilizer treated fields; hence it can be suggested that these plants are facing higher degree of stress compared to the plants cultivated in the fields supplemented with organic fertilizers. A similar pattern was observed in a study on the effect of chemical and bio fertilizers in mustard (Mondal et al. 2017).

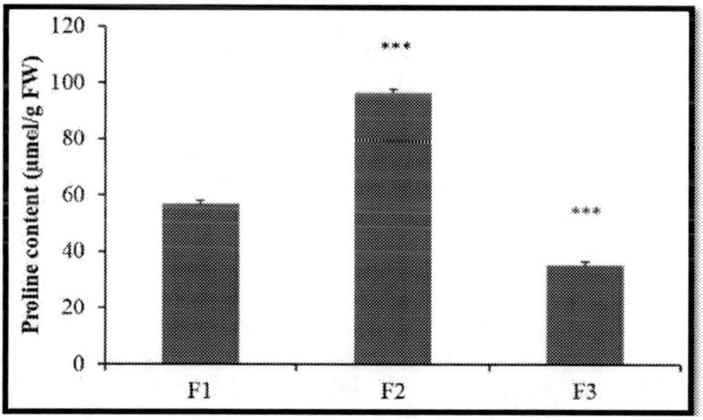

Figure 4. Changes in total proline content in Okra (*Abelmoschusesculentus*). Control group- F1, Chemical fertilizer group- F2, Organic bio fertilizer group-F3. Data are represented as mean ± standard error.
Level of significance: $p < 0.05$ (*); $p < 0.01$ (**); $p < 0.001$(***).

Amount of Lipid Peroxidation

Under stressed condition greater oxidation of lipid occur which leads to Malondialdehyde (MDA) production. To assess the impact of chemical fertilizers the rate of lipid peroxidation was calculated (Figure 5). A similar pattern like that of proline accumulation was observed. MDA content increased significantly in F2 (chemical fertilizer administered) plants (68.26 μmol/L/g FW) compared to that of control F1 plants (42.23 μmol/L/g FW). Level of lipid peroxidation decreased significantly in F3 (organic fertilizer administered) plants (29.13 μmol/L/g FW); the lesser rate of lipid peroxidation points to the fact that organic fertilizers are not inducing any stress on the plant on the contrary to that it is actually having positive impact on plant health. Stress marker such as malonaldehyde (MDA) content was measured as it is a direct indicator of oxidative stress. The findings of the study showed a significantly high MDA accumulation when plants were grown in the field treated with chemical fertilizers. Chemical fertilizers are major contributors of oxidative stress which hastens the rate of lipid peroxidation and hence higher accumulation of MDA. In a recent study it was shown that abiotic stress can induce initial or other subsequent steps of lipid peroxidation and hence enhances MDA production (Olajire & Azeez, 2011). Plants are compelled to produce more and more secondary metabolites under biotic or abiotic stressed condition by ultilising the products of primary metabolism to cope with stress. Hence estimation of major plant secondary metabolites such as polyphenols and flavonoides would give an idea about the extent of impact of stress on the plants which are exposed to high amount of harmful chemical fertilizers.

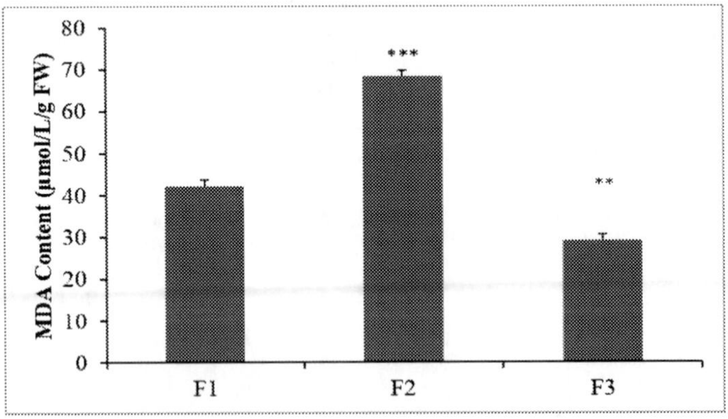

Figure 5. Changes in lipid peroxidation level in Okra (*Abelmoschusesculentus*). Control group- F1, Chemical fertilizer group- F2, Organic bio fertilizer group-F3. Data are represented as mean ± standard error.
Level of significance: p <0.05 (*); p < 0.01 (**); p <0.001(***).

Total Polyphenol Content

Polyphenol content was recorded highest (5935.66 μg GAE/g FW) in the plants cultivated in field supplemented with chemical fertilizers (F3) and least (4438.66 μg GAE/g FW) in plants grown with organic fertilizer supplementation (Figure 6). This significant decrease signifies that organic fertilizers have positive impact on plant health and hence the rate of secondary metabolism has decreased. These findings suggest that plants can withstand stress by enhancing the production of polyphenols because they can inhibit free radical generation (Mishra et al. 2013). Similarly in a study on 25 different varieties of okra it was observed that all the varieties were a good source of natural antioxidants such as polyphenols (Ahiakpa et al. 2013).

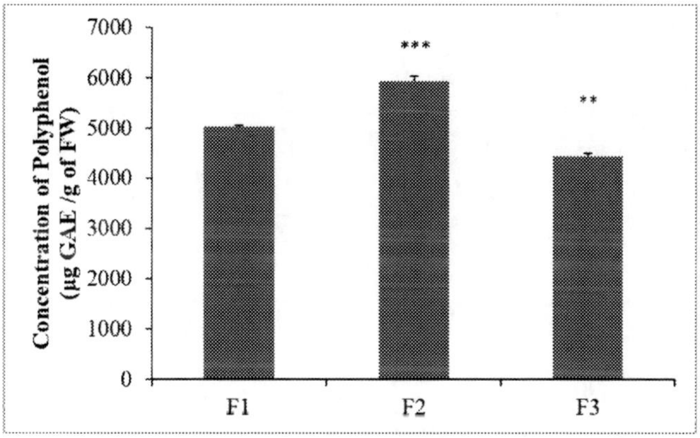

Figure 6. Changes in total polyphenol content in Okra (*Abelmoschusesculentus*).
Control group- F1, Chemical fertilizer group- F2, Organic bio fertilizer group-F3.
Data are represented as mean ± standard error.
Level of significance: $p < 0.05$ (*); $p < 0.01$ (**); $p < 0.001$(***).

Total Flavonoid Content

In a study on metabolomics in okra it was reported that polyphenols and flavonoids are the most abundant secondary metabolites (Xia et al. 2015). Also In this study presence of both polyphenols and flavonoids were recorded. The total flovonoid production has also showed a similar trend like those of polyphenols. Not only it is lesser in F3 plants (organic fertilizer administered) compared to that of chemical fertilizer administered plants but even the amount of flavonoids in F3 plants (742.6 μg RE/g FW) is significantly lesser to that of the control plants (846.67 μg RE/g FW). This observation implies that the organic fertilizers even alleviating the stress hence the accumulation of secondary metabolites are even lesser from that of the control plants (Figure 7). The hydroxyl configuration of the flavonoids enable them to scavenge ROS and other reactive nitrogen species (RNS) as they can donate hydrogen and electron to the radicals to stabilize them (Cao et al. 1997). Altogether these findings and the results of this study portray the adaptive response of okra against chemical fertilizer induced stress by higher production of polyphenols and flavonoids.

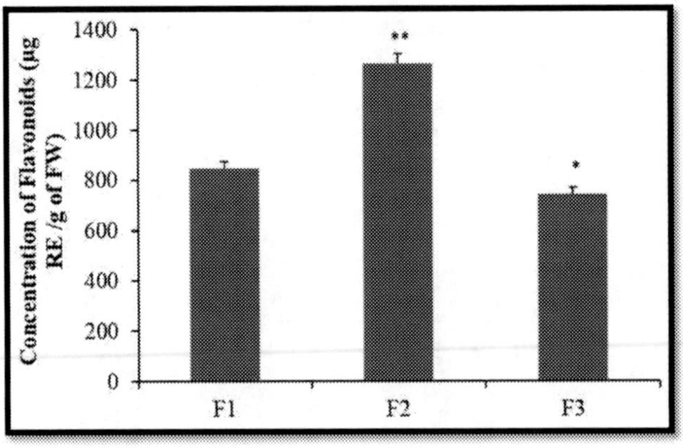

Figure 7. Changes in total flavonoid content in Okra (*Abelmoschusesculentu*).
Control group- F1, Chemical fertilizer group- F2, Organic bio fertilizer group-F3.
Data are represented as mean ± standard error.
Level of significance: $p < 0.05$ (*); $p < 0.01$ (**); $p < 0.001$(***).

Level of DPPH Scavenging

Antioxidative property of polyphenolics is primarily due to their hydroxyl groups and these compounds can also function as scavengers of ROS (Sgherri et al. 2003). A concomitant increase in DPPH activity was recorded along with the increase in both polyphenols and flavonoid content (Figure 8). It is quite obvious as at higher stress level the production of reactive oxygen (ROS) and nitrogen species (RNS) increases and plants try to quench them enhancing the production of secondary metabolites. The least amount of level of DPPH scavenging activity (14%) in plants of F3 group confirms our observation that F3 plants (organic fertilizer administered) are facing least amount of stress among all and hence having minimum stress induced ROS or RNS in them. Phenolic compounds are potential antioxidants as they are capable of quenching ROS and other free radicals (Grubesˇic et al. 2005). Studies have documented tissue specific accumulation of polyphenols occurred under different environmental stresses (Douglas et al. 1996). Hence low lipid peroxidation and DPPH scavenging level in plants grown in organic fertilizer treated fields can be correlated with the low phenolic and flavonoid content in those plants because it indicates that these plants are exposed to lower amount of stress.

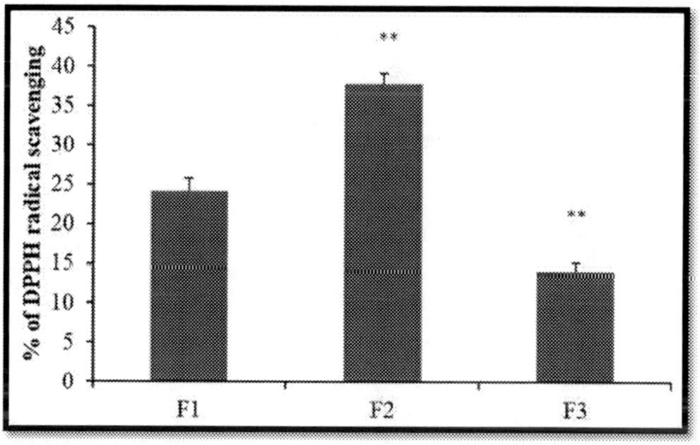

Figure 8. Change inDPPH scavenging activity in Okra(*Abelmoschusesculentus*). Control group- F1, Chemical fertilizer group- F2, Organic bio fertilizer group-F3. Data are represented as mean ± standard error.
Level of significance: p <0.05 (*); p < 0.01 (**); p <0.001(***).

Total Protein Content

It is a marker of plant health. Higher protein content not only signifies that plant is healthy but also indicates that the yield is also good. Among the three groups of plants the group cultivated in the soil supplemented with organic fertilizers had the highest protein content (48.6 mg/g FW) and least (32.2 mg/g FW) in F2 group (chemical fertilizer) (Figure 9). It is quite evident there is almost a 1.5 fold increase in protein content in the plants cultivated in the soil supplemented with organic fertilizers compared to the plants grown in the soil supplemented with chemical fertilizers. It proves that from the nutritional point of view also organic fertilizers are beneficial to the plant.

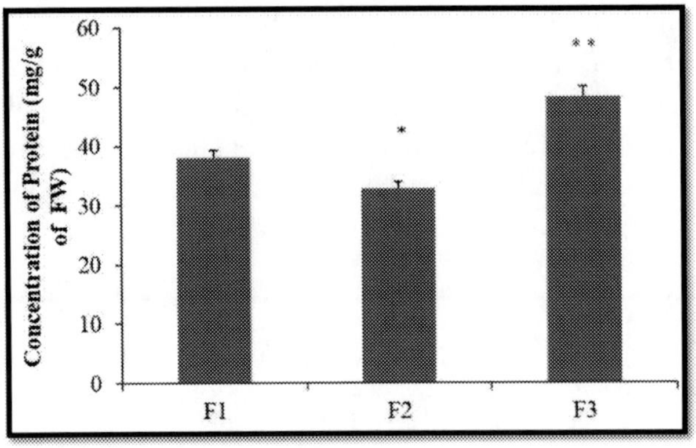

Figure 9. Change in total protein content inOkra (*Abelmoschusesculentus*).
Control group- F1, Chemical fertilizer group- F2, Organic bio fertilizer group-F3.
Data are represented as mean ± standard error.
Level of significance: p <0.05 (*); p < 0.01 (**); p <0.001(***).

Total Carbohydrate Content

A similar pattern like that of protein content was recorded in carbohydrate content also. Plants grown in soil supplemented with organic fertilizer has highest carbohydrate production (17.65 mg/g FW). Use of chemical fertilizer reduced the production of carbohydrates (8.25 mg/g FW) in the plants. Almost more than 2 fold increase in carbohydrate content was recorded in the plants cultivated in the soil supplemented with organic fertilizers compared to the plants grown in the soil supplemented with chemical fertilizers (Figure 10).

In a study on okra it was concluded that organic fertilizer was more potent in enhancing crop yield compared to that of chemical fertilizers but in that study the focus was more on yield and fruit size of okra not on fruit quality or nutrition value (OfosuAnim et al. 2006). In this present study the focus was entirely on the nutritional value and plant health.

In plants cultivated in the field where only organic fertilizer was used showed higher amount of carbohydrate and protein production compared to the control group of plants (no fertilizers used) and plants grown in chemical fertilizer administered fields. This result is congruous with our finding that in these plants (organic fertilizer administered) the level of stress is also lesser compared to that of the other two groups (chemical fertilizer administered plants and the control group of plants without any fertilizer administered respectively) studied.

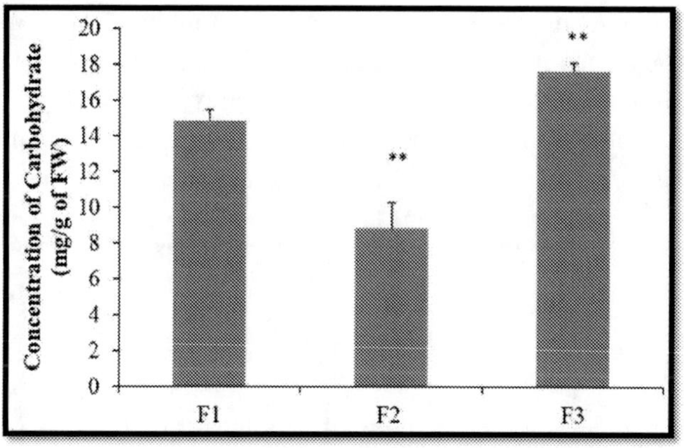

Figure 10. Changes in total carbohydrate content inOkra (*Abelmoschusesculentus*).
Control group- F1, Chemical fertilizer group- F2, Organic bio fertilizer group-F3.
Data are represented as mean ± standard error.
Level of significance: $p < 0.05$ (*); $p < 0.01$ (**); $p < 0.001$(***).

PAL Activity Analysis

Measuring the enzymatic activity of PAL is of utmost importance as it indicates the rate of production of secondary metabolites. In a metabolomics study it was reported higher PAL activity leads to enhanced production of cinnamic acid and other polyphenols such as gallic acid, rutin etc. in okra (Liu et al. 2017). As already discussed plants become more inclined towards secondary metabolism as stress level increases. It was observed that highest PAL activity was recorded in F2 group of plants (chemical fertilizer administered) – 25.78 μM trans cinnamic acid min^{-1} g^{-1} FW; followed by control group (F1) – 11.95 μM trans cinnamic acid min^{-1} g^{-1} FW and lowest in F2 group of plants (organic fertilizer administered) - 7.38 μM trans cinnamic acid min^{-1} g^{-1} FW (Figure 11). This field study reports that there is a significant increase in PAL activity and total polyphenol accumulation in the plants (F2) cultivated in fields where chemical fertilizer was used and a opposite trend was observed in the plants (F3) growing in the land where organic fertilizer was used. This observation clearly implies that the F3 plants are facing least amount of stress hence their PAL activity and rate of secondary metabolism is lesser. The result of this study also shows that in plants grown in organic fertilizer treated fields there was a significant decrease in PAL activity and expression; also total secondary metabolite accumulation was also low.

Studies have also shown that enhancement in PAL activity is coupled with increase in polyphenol production to combat different types of stress (Jouili & Ferjani 2003). Hence it can be propounded from the above observations that organic fertilizers are playing beneficial role in alleviating stress in okra plants hence the level of several stress markers are low compared to that in the plants grown in field administered with chemical fertilizers.

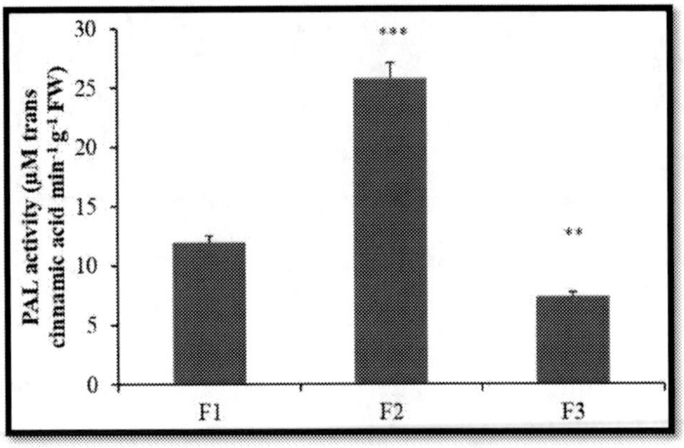

Figure 11. Change of PAL activity in Okra (*Abelmoschusesculentus*).
Control group- F1, Chemical fertilizer group- F2, Organic bio fertilizer group-F3.
Data are represented as mean ± standard error.
Level of significance: $p < 0.05$ (*); $p < 0.01$ (**); $p < 0.001$(***).

Altogether these findings suggest how okra adapts against chemical fertilizer induced stress and toxicity by enhancing the production of natural antioxidants such as polyphenols by inducing the higher production of the main enzyme of phenyl propanoid pathway- Phenylalanine ammonia lyase. The results of this field study proves that from the nutritional point of view also organic fertilizers are beneficial to the plant and hence to the consumers also. Astoundingly the result of using natural organic fertilizers was quite impressive compared to that of chemical fertilizers. Also the use of this biodegradable organic fertilizer is good from point of view of environmental aspect. This is a relatively unexplored domain of plant biological research hence this study would enable us to understand this to some extent the importance of using organic fertilizers from the aspect of plant health and nutritional value of okra. The outcome of this study would encourage us to relinquish the use of chemical fertilizers and use more and more organic fertilizers. A diligent effort towards awareness of using eco friendly biodegradable organic fertilizers supported with data would certainly enable us make people understand the benefit of using organic fertilizers. The outcome of this study clearly suggests that novel organic fertilizer formulated by us not only having beneficial impact on plant health it is also enhancing the nutritional values of okra.

Conclusion

There is a preconceived notion that nutritional value, as well as production of crop and vegetable, can only be increased by applying more chemical fertilizers. The results of this filed study dismiss this wrong proposition. In the present study a detailed comparative assessment between the effect of organic and chemical fertilizers on plant health and nutrition content of the products enable us to conclusively suggest that organic fertilizers are by far more effective in enhancing the nutritional values of okra; besides that, it also alleviates stress and beneficial for the growth and health of the plant. Hence it can be concluded that the use of organic fertilizers, on one hand, reduces the stress level in the plant on the other hand helps the plant to produce important primary metabolites such as protein and carbohydrates which will help plant growth and yield and eventually beneficial for nutritional and health aspects of consumers taking this vegetables on a regular basis.

Acknowledgments

The authors sincerely acknowledge University of Engineering and Management, Kolkata.

References

[1] Lyngdoh, Y. A., Mulge, R. and Shadap, A., 2013. Heterosis and combining ability studies in near homozygous lines of okra [*Abelmoschus esculentus* (L.) Monech] for growth parameters. *The Bioscan.* 8(4), 1275-1279.

[2] Yang, Y., Jin, Z., Mao, P., Jin, J., Huan, J., Yang, M., 2012. Study on anti-fatigue effect of okra extracts. *Chin. J. Mod. Appl. Pharm.* 29, 4.

[3] Mahajan Gupta, R. D., Sharma, R., 2008. Bio-fertilizers-A way to sustainable agriculture. *Agrobios Newsletter.* 6, 36-37.

[4] Hijbeek, R., Ten Berge, H. F. M., Whitmore, A. P., Barkusky, D., Schroder, J. J., Van Ittersum, M. K., 2018. Nitrogen fertilizer replacement values for organic amendments appear to increase with N application rates. *Nutrient Cycling in Agroecosystems.* 110(1), 105-115.

[5] Nelson, A. R. L. E., Ravichandran, K., Antony, U., 2019. The impact of the Green Revolution on indigenous crops of India. *Journal of Ethnic Foods.* 6, 8. https://doi.org/10.1186/s42779-019-0011-9.

[6] Yamaguchi, R., Hwang L. S., 2015. Food safety program in asian countries. *J Nutr Sci Vitaminol* (Tokyo) 61 Suppl: S53-54.

[7] Sharma, N., Singhvi, R., 2017. Effects of chemical fertilizers and pesticides on human health and environment: a review. *International Journal of Agriculture, Environment and Biotechnology.* 10(6), 675-679. 10.5958/2230-732X.2017.00083.3.

[8] Mal, B., Mahapatra, P., Mohanty, S., Mishra, N., 2013. Growth and yield parameters of okra (Abelmoschus esculentus) influenced by Diazotrophs and chemical fertilizers. *J. Crop and Weed.* 9(2), 109-112.

[9] Bano R., Soni A. K., Singh J., Singh S. P. and Mahawar A. K. (2018) Efficacy of organic manures and bio-fertilizers on yield attributes and yield of okra *var.* Chameli. *International Journal of Current Microbiology and Applied Sciences* 7(9): 3668-3675.

[10] Liang, N., Kitts, D. D., 2016. Role of chlorogenic acids in controlling oxidative and inflammatory stress conditions. *Nutrients,* 8:16.

[11] Goujon, T., Sibout, R., Eudes, A., MacKay, J., Jouanin, L., 2003. Genes involved in the biosynthesis of lignin precursors in *Arabidopsis thaliana. Plant Physiology and Biochemistry.* 41, 677–687.

[12] Dixon, R. A., Paiva, N., 1995. Stress-induced phenylpropanoid metabolism. *Plant Cell* 7, 1085-1097.

[13] Mittler, R., 2002. Oxidative stress, antioxidants and stress tolerance. *Trends in Plant Science* 7.9, 405-410.

[14] Ahiakpa K., Quartey, E. K., Achel, G., Achoribo, E., 2013. Total flavonoid, phenolic contents and antioxidant scavenging activity in 25 accessions of okra (*Abelmoschus* spp L.). *African Journal of Food Science and Technology.* 4(5), 129-135.

[15] Lichtenthaler, H. K., 1987. Chlorophylls and carotenoids: pigments of photosynthetic biomembranes. *Methods in Enzymology.* 148, 350-382.

[16] Bates, L., Waldren, R. P., Teare, I. D., 1973. Rapid determination of free proline for water-stress studies. *Plant and Soil.* 39, 205–207.

[17] Heath, R. L., Packer, L., 1968. Photoperoxidation in isolated chloroplasts: I. Kinetics and stoichiometry of fatty acid peroxidation. *Archives in Biochemistry and Biophysics.* 125, 189-198.

[18] Brolis, M., Gabetta, B., Fuzzati, N., Pace, R., Panzeri, F., Peterlongo, F., 1998. Identification by high-performance liquid chromatography–diode array detection–mass spectrometry and quantification by high performance liquid chromatography–

UV absorbance detection of active constituents of *Hypericum perforatum. Journal of Chromatography.* 825, 9–16.

[19] Singleton, V. L., Orthofer, R., Lamuela-Raventos, R. M., 1999. Analysis of total phenols and other oxidation substrates and antioxidants by means of folin-ciocalteu reagent. *Methods in Enzymology.* 299, 152-178.

[20] Talukder, P., Talapatra, S., Ghoshal, N., Sen Raychaudhuri, S., 2016. Antioxidant activity and high-performance liquid chromatographic analysis of phenolic compounds during *in vitro* callus culture of *Plantago ovata* Forsk and effect of exogenous additives on accumulation of phenolic compounds. *Journal of the Science of Food and Agriculture.* 96(1), 232-244.

[21] Lin, J. Y., Tang, C. Y., 2007. Determination of total phenolic and flavonoid contents in selected fruits and vegetables, as well as their stimulatory effects on mouse splenocyte proliferation. *Food Chemistry.* 101, 140-147.

[22] Brand-Williams, W., Cuvelier, M. E., Berset, C., 1995. Use of a free radical method to evaluate antioxidant activity. *Lebensmittel-Wissenschaft & Technologie.* 28, 25-30.

[23] Dos Santos, W. D., Ferrarese, M. L. L., Finger, A., Teixeira, C. A. N., Ferrarese-Filho, O., 2004. Lignification and related enzymes in *Glycine max* root growth-inhibition by ferulic acid. *Journal of Chemical Ecology.* 30, 1203-121.

[24] Kovacik, J., Klejdus, B., Backor, M., Repcak, M., 2007. Phenylalanine ammonia-lyase activity and phenolic compounds accumulation in nitrogen-deficient *Matricaria chamomilla* leaf rosettes. *Plant Science.* 172, 393-399.

[25] Edwards, K., Johnstone, C., Thompson, C., 1991. A simple and rapid method for the preparation of plant genomic DNA for PCR analysis. *Nucleic Acids Research.* 19(6), 13.

[26] Okutucu, B., 2017. Comparison of five methods for determination of total plasma protein concentration. *Journal of Biochemical and Biophysical Methods.* 70(5), 709-711.

[27] Hedge, J. E., Hofreiter, B. T., 1962. *Carbohydrate Chemistry* 17. Whistler, R. L. and Be Miller, J. N., Eds., Academic Press, New York.

[28] Tiamiyu, R. A., Ahmed, H. G., Muhammad, A. S. 2012. Effect of sources of organic manure on growth and yields of okra (*Abelmoschus esculentus L.*) in Sokoto, Nigeria. *Nigerian Journal of Basic and Applied Science.* 20(3), 213-216.

[29] Agarwal, S., Pandey, V., 2004. Antioxidant enzyme responses to NaCl stress in *Cassia angustifolia. Biologia Plantarum.* 48, 555-560.

[30] Mondal, T., Datta, J. K., Mondal, N. K., 2017. Chemical fertilizer in conjunction with biofertilizer and vermicompost induced changes in morpho-physiological and bio-chemical traits of mustard crop. *Journal of the Saudi Society of Agricultural Sciences.* 16, 135–144.

[31] Olajire, A. A., Azeez, L., 2011. Total antioxidant activity, phenolic, flavonoid and ascorbic acid contents of Nigerian vegetables. *African Journal of Food Science and Technology.* 2, 22-29.

[32] Mishra, A., Sharma, A. K., Kumar, S., Saxena, A. K., Pandey, A. K., 2013. *Bauhinia variegata* leaf extracts exhibit considerable antibacterial, antioxidant and anticancer

activities. *Bio Med Research International,* vol. 2013, Article ID 915436, 10 pages. 2013:915436. doi: 10.1155/2013/915436.

[33] Xia, F., Zhong, Y., Li, M., Chang, Q., Liao, Y., Liu, X., Pan, R., 2015. Antioxidant and anti-fatigue constituents of okra. *Nutrients.* 7, 8846–8858.

[34] Cao, G., Sofic, E., Prior, R. L., 1997. Antioxidant and prooxidant behavior of flavonoids: structure-activity relationships. *Free Radical Biology and Medicine.* 22 (5), 749-760.

[35] Sgherri. C., Cosi, E., Naveri-izzo, F., 2003. Phenols and antioxidative status of *Raphanussativus* grown in copper excess. *Physiol. Plant.* 118, 21-28.

[36] Grubesic, R. J., Vukovic, J., Kremer, D., Vladimir-Knězevic, S., 2005. Spectrophotometric method for polyphenols analysis: Prevalidation and application on *Plantago* L. species. *Journal of Pharmaceutical and Biomedical Analysism.* 39, 837-842.

[37] Douglas, C. J., 1996. Phenylpropanoid metabolism and lignin biosynthesis: from weeds to trees. *Trends in Plant Science.* 1, 171-178.

[38] OfosuAnim, J., Blay, E. T., Frempong, M. E., 2006. Effects of organic manure on okra (*Abelmoschus esculentus* (L.) Moench) production. *Journal of Applied Horticulture.* 8(2), 155-158.

[39] Liu, J., Yuana, Y., Wua, Q., Zhaoa, Y., Jianga, Y., Johna, A., Wena, L., Lia, T., Jiana, Q., Yanga, B., 2017. Analyses of quality and metabolites levels of okra during postharvest senescence by 1H-high resolution NMR. *Postharvest Biology and Technology.* 132, 171–178.

[40] Jouili, H., Ferjani, E. E., 2003. Changes in antioxidative and lignifying enzyme activities in sunflower roots (*Helianthus annus* L.) stressed with copper excess. *Compt. Rend. Biol.* 326, 639-644.

Chapter 10

Preparation and Evaluation of Functional Properties of Tannic Acid Loaded Corn Starch Film

Raj Kumar[1], Gargi Ghoshal[1], Meenakshi Goya[1] and Shashikant Yadav[2,*]

[1]Dr. S. S. Bhatnagar University Institute
of Chemical Engineering & Technology,
Panjab University, Chandigarh, India
[2]Department of Chemical Engineering,
Dr. B. R. Ambedkar National Institute of Technology,
Jalandhar, Punjab, India

Abstract

The main focus of this work was to develop biodegradable films using corn starch and tannic acid and to investigate the effect of tannic acid on the functional properties such as thickness, swelling index, moisture content, opacity, tensile strength, and elongation of the film. The incorporation of tannic acid in various concentrations (1, 3, 5, and 7% w/w of starch) into the corn starch improved the moisture permeability, tensile strength, swelling index, and solubility of the film. The modified starch film also decreases the elongation at breakpoint and opacity. FTIR analysis of the film confirmed the establishment of new interactions between tannic acid and starch. It was also observed that the strong interactions among the starch chains developed through hydrogen bonding. In conclusion, adding tannic acid in the corn starch offered an excellent way to improve the functional properties of the film.

---

[*] Corresponding Author's Email: shashikant529@gmail.com.

In: Proceedings of BIOSPECTRUM
Editors: Sanket J. Joshi, Susmita Mukherjee and Moupriya Nag
ISBN: 978-1-68507-985-7

Keywords: corn starch, tannic acid, FTIR, mechanical property

Introduction

Natural polysaccharides such as a variety of starch and their derivatives, cellulose and their derivatives, and proteins are widely used for the preparation of biodegradable films and coatings. Biodegradable films and coatings made from natural polymers have attracted the continued attention of academicians and researchers due to their wide applications in food and pharmaceutical packaging (Shah et al., 2016; Elsabee and Abdou 2013).

Starch can be obtained from various natural resources to develop biodegradable films especially for the food industry. These films protect foods from physical damages, moisture, light and gas, thus increasing food integrity and shelf life (Dhall 2016). Among all polysaccharides, the use of starch to prepare films and coatings is advantageous due to its large availability, low cost, biodegradability, non-toxic, and excellent film-forming nature, which makes starch one of the most attractive raw material for preparation of films or coatings (Caz 2017). However, inferior functional properties such as moisture migration barrier and mechanical properties restrict their usages in the food packaging (Basiak, Lenart, and Debeaufort 2017). To address these limitations, physical and chemical modification techniques are used to modify the starch film properties. For example, several cross-linking agents and natural fillers are adopted to enhance the functional properties of films. In the physical modification process, sometimes non-compatibility among the polymers and insufficient improvement in the functional properties limit their use in food packaging.

To address these problems, chemical modification techniques are used in which cross-linking agents such as formaldehyde, boric acid, oxidize sucrose, and tannic acid improve the functional properties of the films (Zhang et al., 2017; Yin et al., 2005; Xu et al., 2015). However, some cross-linking agents are toxic in nature, and therefore, not recommended in the food industry (Kieboom 2005).

To overcome these problems, the tannic acid derived from plants can be used as a cross-linking agent. Tannic acid exhibits excellent cross-linking ability due to its multi-carboxyl groups in the chemical formula (Rivero, García, and Pinotti 2010). Therefore, the functional properties of starch films can be significantly improved by the addition of tannic acid. Limited work has been done on tannic acid/starch films. The present research focused on developing modified corn starch-based films using the casting method. It examined the effect of tannic acid on the functional properties of corn starch films plasticized with sorbitol.

Materials and Method

All the chemicals purchased were of analytical grade. Corn starch, tannic acid, and calcium chloride were purchased from Loba chem (Mumbai, India). Sorbitol was purchased from HiMedia (Mumbai, India).

Preparation of Films

Modified corn starch films were prepared using the casting method. In brief, an aqueous suspension of corn starch was synthesized by dissolving 3 g starch, 0.75 g sorbitol, and tannic acid (1, 3, 5, and 7% w/w of total starch) in 100 mL distilled water.

The solution was mixed uniformly using a magnetic stirrer hot plate at 80 ± 1°C for 40 min. Thereafter, the solutions were homogenized using a high speed homogenizer at 12000 rpm for 2 min. After that, the solutions were degassed using a vacuum oven. These homogeneous modified starch solutions with no air bubbles were evenly spread on casting plates and dried at room temperature. The dried films were peeled off from the plates and stored in air-tight bags.

Characterization

Film Thickness

Variations in the film thickness with different concentrations of tannic acid in the films were recorded using a digital micrometer with an accuracy of ± 0.001 mm (Mitutoyo, Japan).

Moisture Content

The gravimetric method was used to determine the moisture content (MC) of films.

MC was determined according to equation (1):

$$MC(\%) = \frac{m_a - m_b}{m_a} * 100 \quad (1)$$

Swelling Index

The weight of dried samples (105 ± 1°C for 24 h) was recorded, and the samples were then immersed in 25 mL double distilled water at 25 ± 1°C for 2 min. Swollen samples were removed and weighed again.

The swelling index (SI %) of samples was calculated using equation (2):

$$SI(\%) = \frac{M_a - M_b}{M_a} * 100 \quad (2)$$

where M_a and M_b are the weights of the swelled samples and dried samples, respectively.

Solubility

The water solubility of films was measured using the reported method (Matta et al., 2015).

Water Vapor Barrier Property

Moisture migration permeability was evaluated using gravimetric modified beaker technique according to ASTME method 96/E96M (2014).

Opacity

The appearance and film opacity were measured using the Hunter Lab Colorimeter (Color Flex EZ, Hunter Lab, USA). Before starting the measurement, the instrument was calibrated with a white and a black background. Films opacity (OP %) was determined using equation (3):

$$OP(\%) = \frac{OP_a}{OP_b} * 100 \qquad (3)$$

where: OP_a and OP_b are the opacity of the films with white and black backgrounds, respectively.

Mechanical Property

Films elongation (E %) and film strength (FS) were calculated using Texture Analyzer (TA.XT Plus, Stable Microsystems) and according to ASTM standard method 828-97 (2002).

FTIR Analysis

The interactions among starch chains were identified using FTIR spectra (Shimadzu -8400, Japan). The spectra were acquired in the scanning range of 500 to 4000 cm^{-1} with a resolution of 4 cm.

Statistics

Thickness, swelling index, moisture content, solubility, moisture migration permeability, and mechanical properties were determined in triplicate. The results were analyzed as mean ±SD and studied using Graph Pad 6.0 software. Analysis of variance determined by ANOVA method ($P < 0.05$).

Result and Discussion

Thickness

The change in thickness of starch films was slightly varied with tannic acid, as shown in Figure 1(A). Addition of tannic acid to starch films increased dry mass of the films thereby increased the thickness of the films.

Moisture Content

The effect of adding tannic acid on the moisture content of the films is shown in Figure 1(A). The addition of tannic acid into the films significantly decreased ($p \leq 0.05$) the moisture content of the films. The moisture content was decreased due to esterification and hydrogen bonding among starch chains (Masamba et al., 2016; Menzel et al., 2013).

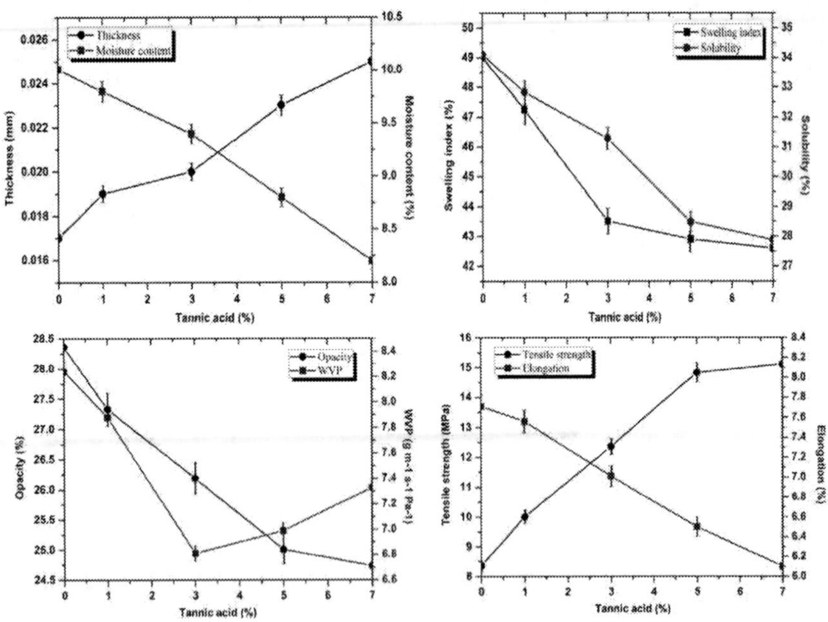

Figure 1. Effects of tannic acid on the functional properties (1, 3, 5 & 7% w/w) thickness, moisture content (A), solubility, swelling index (B), opacity, WVP (C), tensile strength & elongation (D).

Swelling Index

The variation in swelling index by adding tannic acid to the film is shown in Figure 1(B). The swelling index of the films decreased considerably ($p \leq 0.05$) with the addition of tannic acid. The hydrophilic sites were reduced after tannic acid was added to the films. Thus the interactions of water molecules with the film were decreased, leading to a decrease in SI (Ma et al., 2017; Cao, Fu, and He 2007).

Solubility

Films with high solubility are not acceptable for use in food packaging. Therefore, it is necessary to optimize the solubility of the films. The variation in water solubility with tannic acid is demonstrated in Figure 1(B). As tannic acid concentration in films increases, the solubility of the films decreases significantly ($p \leq 0.05$). The decrease in the solubility can be expressed by the fact that in the starch films tannic acid created cross-linking and hydrogen bonding among the starch chains. As a result, a compact and interlocked starch matrix was developed. Thus, the water affinity of the films toward water decreases, resulting in a decrease in solubility (Hager, Vallons, and Arendt 2012; Masamba et al., 2016).

Water Vapor Permeability

Films with high moisture permeability are not suitable for packaging as it creates serious issues such as changes in the texture integrity of films when moisture is gained or lost through the environment. The addition of tannic acid to the films significantly reduced the moisture permeability of the films ($p \leq 0.05$), as shown in Figure 1(C). The decrease in moisture permeability is due to the fact that the tannic acid in the films increased the tortuous moisture path length and the interaction among starch chains; thus, the moisture migration path evolved in the starch matrix. Therefore, water vapor required to travel a greater distance in the films, which reduces the water vapor migration rate. The results of solubility and swelling index supported the reduction of the water vapor permeability of films (Menzel et al., 2013).

Opacity

Food packaging films with sufficient opacity protect light-sensitive food and help to improve the shelf life of foods. Therefore, packaging films with high opacity are suitable for light-sensitive foods and pharmaceutical materials. From the findings, it was observed that as the concentration of tannic acid increased, the opacity of the films decreased significantly ($p \leq 0.05$), as shown in Figure 1(C). Strong interactions among the starch chains through the esterification process reduced the crystallinity of starch, thereby reducing light penetration through the films (Seligra et al., 2016).

Mechanical Property

Food packaging provides integrity against physical damages that may occur during food transportation and distribution. The film strength and elongation significantly ($p \leq 0.05$) increased in the film with tannic acid at various concentrations, as shown in Figure 1(D). Therefore, the films exhibit improved mechanical property. The increased interfacial interactions between the chains resulted in decreased mobility between the chains, which improved the film strength through forming cross-linking between the starch chains. In contrast, elongation was decreased (Hager, Vallons, and Arendt 2012).

FTIR Analysis

Interactions in the starch chains were created via hydrophobic ester bonding through esterification reaction between the functional groups of starch, and tannic acid was observed using FTIR technique as displayed in Figure 2(A&B). The appearance of peaks of native starch films and modified corn starch spectrums at 3289.51 cm^{-1} and 3266.23 cm^{-1}, respectively, address the stretching vibration of hydroxyl groups in the starch matrix and have more freedom to vibrate and make hydrogen bonding with water (Shi et al., 2007).

Both modified and native starch films showed peaks around 1630 cm^{-1}, indicating the swelling of starch molecules with water gain. Other peaks in both the modified starch and native starch in the range of 1000-1200 cm^{-1} represented hydrolysis in C-O and C-O-C groups. In the modified starch spectrum, the peak at 1751.89 cm^{-1} addressed C = O stretch vibration, indicating cross-linking among the starch matrix via esterification reaction (Shi et al., 2007; Seligra et al., 2016).

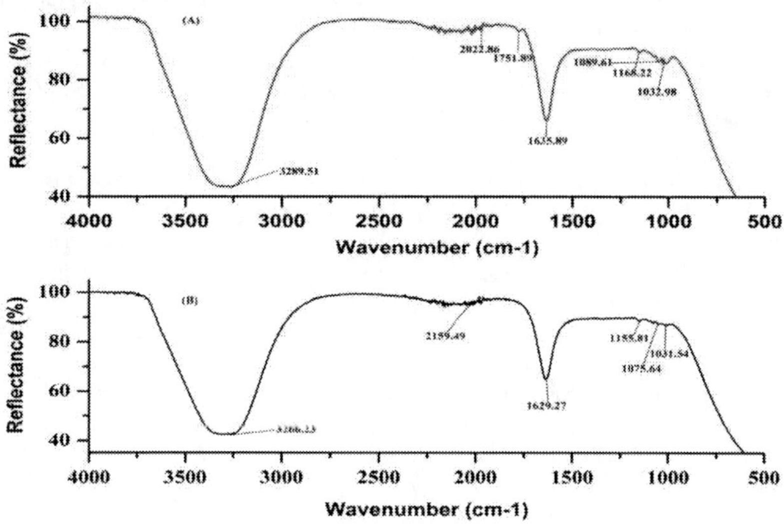

Figure 2. FTIR spectra of modified corn starch (A) and native corn starch (B).

Conclusion

In this work, cross-linking among starch chains was successfully achieved through esterification using plant-based tannic acid, which modified the functional properties of corn starch films. However, the high acid content in the starch matrix probably increased free space volume between starch chains, leading to increased moisture migration rate and elongation and decreased film strength. FTIR analysis confirmed the cross-linking among the starch matrix. In conclusion, packaging films made with modified corn starch using casting method showed superior functional properties and suggested its applications in food packaging and coating.

References

Basiak, E., Andrzej Lenart, and Frédéric Debeaufort. 2017. "Effect of Starch Type on the Physico-Chemical Properties of Edible Films." *International Journal of Biological Macromolecules* 98: 348-56. https://doi.org/10.1016/j.ijbiomac.2017.01.122.

Cao, N., Yuhua Fu, and Junhui He. 2007. "Mechanical Properties of Gelatin Films Cross-Linked, Respectively, by Ferulic Acid and Tannin Acid." *Food Hydrocolloids* 21 (4): 575-84. https://doi.org/10.1016/j.foodhyd.2006.07.001.

Caz, P. 2017. "Food Hydrocolloids Polysaccharide-Based Fi Lms and Coatings for Food Packaging: A Review" *Food Hydrocolloids* 68: 136-48.

Dhall, R. K. 2016. "Advances in Edible Coatings for Fresh Fruits and Vegetables : A Review Advances in Edible Coatings for Fresh Fruits and Vegetables." *Critical Reviews in Food Science and Nutrition* 8398 (January). https://doi.org/10.1080/10408398.2010.541568.

Elsabee, M. Z., and Entsar S. Abdou. 2013. "Chitosan Based Edible Fi Lms and Coatings: A Review" *Mater Sci Eng C Mater Biol Appl* 33: 1819-41.

Hager, A. S., K. J. Vallons, and E. K. Arendt. 2012. "Influence of Gallic Acid and Tannic Acid on the Mechanical and Barrier Properties of Wheat Gluten Films." *J Agric Food Chem* 60 (24): 6157-63. https://doi.org/10.1021/jf300983m.

Kieboom, T. 2005. "Glutaraldehyde Cross-Link Analogues From" *Starch* 57: 161-65. https://doi. org/10.1002/star.200400357.

Ma, M., Suozhuai Dong, Munir Hussain, and Wenlong Zhou. 2017. "Effects of Addition of Condensed Tannin on the Structure and Properties of Silk Fibroin Film." *Polymer International* 66 (1): 151-59. https://doi.org/10.1002/pi.5272.

Masamba, K., Yue Li, Joseph Hategekimana, Fei Liu, and Jianguo Ma. 2016. "Effect of Gallic Acid on Mechanical and Water Barrier Properties of Zein-Oleic Acid Composite Films" *Journal of Food Science and Technology* 53 (May): 2227-35. https://doi.org/10.1007/s13197-015-2167-7.

Matta, F., Silvia Maria, Thiago Caon, José Ignacio, Lucia Helena, Innocentini Mei, Farayde Matta Fakhouri, et al., 2015. "Edible Films and Coatings Based on Starch/Gelatin: Film Properties and Effect of Coatings on Quality of Refrigerated Red Crimson Grapes." *Postharvest Biology and Technology* 109: 57-64. https://doi.org/10.1016/j.postharvbio.2015.05.015.

Menzel, C., Erik Olsson, Tomás S. Plivelic, Roger Andersson, Caisa Johansson, Ramune Kuktaite, Lars Järnströmb, and Kristine Koch. 2013. "Molecular Structure of Citric Acid Cross-Linked Starch Films." *Carbohydrate Polymers* 96 (1): 270-76. https://doi.org/10.1016/j.carbpol.2013.03.044.

Rivero, S., M. A. García, and A. Pinotti. 2010. "Crosslinking Capacity of Tannic Acid in Plasticized Chitosan Films" *Carbohydrate Polymers* 82: 270-76. https://doi.org/10.1016/j.carbpol.2010.04. 048.

Seligra, P. G., Carolina Medina Jaramillo, Lucía Famá, and Silvia Goyanes. 2016. "Biodegradable and Non-Retrogradable Eco-Films Based on Starch-Glycerol with Citric Acid as Crosslinking Agent." *Carbohydrate Polymers* 138: 66-74. https://doi.org/10.1016/j.carbpol.2015.11.041.

Shah, U., Farah Naqash, Adil Gani, and F A Masoodi. 2016. "Art and Science behind Modified Starch Edible Films and Coatings: A Review" *Comprehensive Reviews in Food Science and Safety* 15: 568-80. https://doi.org/10.1111/1541-4337.12197.

Shi, R., Zizheng Zhang, Quanyong Liu, Yanming Han, Liqun Zhang, Dafu Chen, and Wei Tian. 2007. "Characterization of Citric Acid/Glycerol Co-Plasticized Thermoplastic

Starch Prepared by Melt Blending." *Carbohydrate Polymers* 69 (4): 748-55. https://doi.org/10.1016/j.carbpol.2007.02.010.

Xu, H., Hazal Canisag, Bingnan Mu, and Yiqi Yang. 2015. "Robust and Flexible Films from 100% Starch Cross-Linked by Biobased Disaccharide Derivative." *ACS Sustainable Chemistry and Engineering* 3 (11): 2631-39. https://doi.org/10.1021/acssuschemeng.5b00353.

Yin, Y., Jianfang Li, Yingchun Liu, and Zhong Li. 2005. "Starch Crosslinked with Poly (Vinyl Alcohol) by Boric Acid" *Journal of Applied Polymer Science* 96 (September 2004): 1394-97. https://doi.org/10.10 02/app.21569.

Zhang, Y., Jiayi Zhu, Hongbo Ren, Yutie Bi, and Lin Zhang. 2017. "Synthesis and Properties of Melamine-Starch Hybrid Aerogels Cross-Linked with Formaldehyde." *Journal of Sol-Gel Science and Technology* 83 (1): 44-52. https://doi.org/10.1007/s10971-017-4375-2.

Chapter 11

Enhanced Production of *Citrus macroptera*: A Nutrition Rich Fruit Crop Plant

Susmita Shukla* **and Ritupriya Singh**
Applied Plant Biotechnology Research Lab,
Centre for Plant and Environmental Biotech,
Amity Institute of Biotechnology,
Amity University, Noida (U.P), India

Abstract

According to the Botanical Survey of India (BSI), 10% of the plants in the Nokrek reserve are rare and endangered and few indigenous economically and commercially important species of citrus have been listed as endangered by The World Conservation Union due to enhanced human intervention around their natural habitat. *C. macroptera* is among one of them whose natural population is diminishing at an alarming rate due to overexploitation. This species is medicinally important duc to its antioxidant properties and is commercially utilized because of the flavoring properties of its peel thus, there is a dire need for developing strategies to conserve, manage and utilize the plant species through plant tissue culture technique. The present study deals with the efforts made to establish an efficient protocol for *C. macroptera* through micropropagation technique using nodal explants from in vitro grown seedlings. The seeds were surface sterilized using various sterilants and the most suitable treatment was used further for the selection of explants. Effect of different types of plant tissue culture media on the percentage seed germination and the initiation of cultures for shoot proliferation was evaluated. The culture media was supplemented with different

---

* Corresponding Author's Email: sshukla3@amity.edu.

In: Proceedings of BIOSPECTRUM
Editors: Sanket J. Joshi, Susmita Mukherjee and Moupriya Nag
ISBN: 978-1-68507-985-7

concentrations of phytohormones and adjuvants. The cultures were initiated, and multiple shoot proliferation was obtained.

Keywords: *C. macroptera*, micropropagation, sterilization, multiplication, endangered

Introduction

Citrus (family Rutaceae) is one of the most consumed fruits of economical and commercial importance as well as has nutritional and medicinal properties. The production is native to a large area, extending from the North-eastern part of the Indian foothills of the Himalayas to China, Burma, Thailand. Globally, about 60% of the world flora is estimated to be under protection according to the Global Strategy for Plant Conservation (Coates & Dixion, 2007). One approach to ex-situ conservation is the use of the tools of plant biotechnology. This combines techniques that use plant DNA, cells, tissues, and organs, to provide viable options for ex-situ conservation of plant germplasm (Chen & Chen, 2007; Parawira & Khosa, 2009).

Plant biotechnology can be used to rapidly multiply and increase populations of threatened species through micropropagation (Bitters & Murashige 1967; Hamza et al., 2013; Debergh et al., 1990). Citrus is the third largest fruit crop after Banana and Mango in terms of area under cultivation. The yield of citrus fruits in developed countries such as Brazil, the USA, and Indonesia is likely more (22-35t/ha) as compared to India which is quite low (8.8 t/ha). *Citrus macroptera* Montr. is known to be one of the endangered semi-wild species of citrus native to the northeastern part of India. Reportedly, this species grows in the North Cacher, Karbi Anglong district of Assam, Manipur, Mizoram, Khasi Hills of Meghalaya, and Tripura (Tanaka 1954; Bhattacharya and Dutta 1956; Swingle and Reece 1967; Sharma et al., 2004). *C. macroptera* is a multipurpose tree with rich nutritional and medicinal content. It is used by the tribal people of the Northeast for the treatment of various diseases due to which the Citrus diversity has declined at a rapid rate (Malik et al., 2006).

The employment of an advanced biotechnological approach such as plant tissue culture has great potential for a very high multiplication rate to produce thousands of plants in a short span of time and could be propitious to conserve threatened and endangered plant species(Bunn et al., 2007; Button & Kochba

1977; Almeida et al., 2006; Jona & Menini,1987). Various protocols for the micropropagation of citrus species have been developed (Prodhan et al.,2016; Ling & Iwamasa, 1997; Hussey 1977; Engelmann 1991). The in vitro regeneration protocol has been reported for shoot organogenesis protocol for *C. indica* Tanaka, and (60%) of the rooted plantlets were acclimatized successfully (Laskar et al., 2009) . (Normah et al., 1997) established a successful system for *C. halimii* an endangered species of southeast Asia, and (83.3%) of the rooted plantlets survived. An efficient protocol by (Haripyaree et al., 2011) was developed for *C. megaloxycarpa* Lush, using shoot tip culture, and about 120 plantlets were regenerated from a single seedling. In the case of *C. macroptera*, an attempt to develop an initiation protocol has been done by (Shukla et al., 2018) and on average, (4) micro shoots were produced from the nodes collected through in-vitro grown seedling in the combination of MS medium supplemented with BAP (1mg/L).

Further efforts are being made to establish a full protocol. With these findings, an attempt has been made for the development of a system for the initiation and multiplication of shoots of *C. macroptera* from the explants collected from the mature plant (ex-vitro nodal sections).

Methods

Plant Materials

Ex vitro nodes were taken as the starting material for setting up the experiment. *Ex vitro* nodes were collected from the field. 1cm long nodal segments were surface sterilized by washing the node thoroughly under running tap water for 10 minutes followed by 1% Bavistin. The nodes were washed again and the treatment with 2 to 3 drops of Tween 20(detergent) was given. Under aseptic conditions, nodes were treated with 5% NaOCl for 3 minutes and rinsed thrice with autoclaved water followed with the treatment of 0.1% $HgCl_2$, then washed with 70% Ethanol for 1 minute and finally rinsed with autoclaved water thoroughly before inoculation. After surface sterilization, nodes were inoculated in different culture media.

Media Compositions

To induce the initiation of bud break and shoot proliferation different media compositions (MS, WPM, B5, and MT) alone as a control and in combination supplemented with BAP (1 mg/L) were used.

Optimum Conditions

The inoculated cultures were incubated in the culture room. The temperature of the culture room has been maintained at 25 ± 2 °C under 16/8 h (light/dark) photoperiod. Every reading was taken at regular intervals.

Data Analysis

All experiments were repeated three times with five replicates for each experiment. Each nodal section was inoculated in 10 ml of media in culture tubes and 15 replicons were given for each variant. The data were statistically analyzed using one-way analysis of variance (ANOVA) and the significance value was taken at $\alpha = 0.05$.

Result and Discussion

The ex-vitro nodes collected from the field were inoculated and minimum contamination was found when the nodes were treated with the 5% NaOCl and 0.1% $HgCl_2$ the result in Table 1 suggested that almost 100% of nodes showed the initiation of bud break and micro-shoots formation in the different plant tissue culture basal media (WPM, MS, MT & B5) and in combination with different concentrations of BAP as suggested in the findings of (Miah et al., 2008) and the best suitable results were found in the MS medium supplemented with BAP(1mg/L) whereas in the B5 medium after a certain period of time, the plant tissues started showing wilting and necrosis started. WPM and MT media responded with poor shoot proliferation. The nodal segment showed profuse growth and was initiated successfully. Initially, it was difficult to separate the micro-shoot and get the nodes for sub-culturing as the distance between the nodes of micro-shoots were very less therefore sub-culturing of the intact micro-shoots were transferred into fresh medium for subsequent two sub-culturing cycle thereafter nodes of the micro shoot were inoculated from the third sub-culture as mentioned in Table 2, on an average maximum of 5.8 micro- shoots were obtained till the 6th sub-culture and a decline in the number of micro-shoots as well as the length were observed in the 7th sub-culture stage. The length of the micro-shoot obtained was very small as compared to other micro-propagated Citrus species. (Laskar et al., 2009) reported a maximum of 4.63 cm shoot length of the micro-shoots derived from the callus of *C. indica*. So, further, the experiments were carried out with the different concentrations of Gibberellic acid (0.0, 0.1, 0.5 & 1.0 mg/L) for the elongation of shoots and the best result was obtained with the 0.5mg/L GA whereas the higher concentration (1.0mg/L) started showing callus formation in the explants. The maximum shoot elongation after supplementing GA was 2.02 ± 0.03 cm which was found suitable for the sub-culturing experiment. The result indicates that the single shoot subculturing is possible when the shoot length increases to 2cm long after being fortified with GA which is required to obtain the best suitable result for further experiments to be carried out.

Figure 1. Initiation of bud break.

Figure 2. Multiple micro-shoots.

Figure 3. Multiple shoot formation.

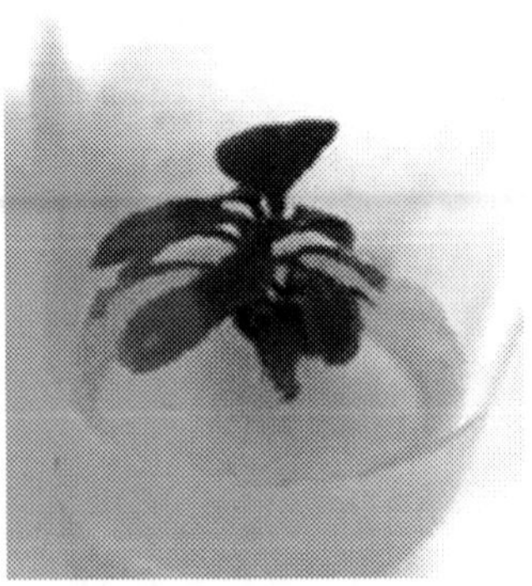

Figure 4. Shoot elongation (GA_3) formation.

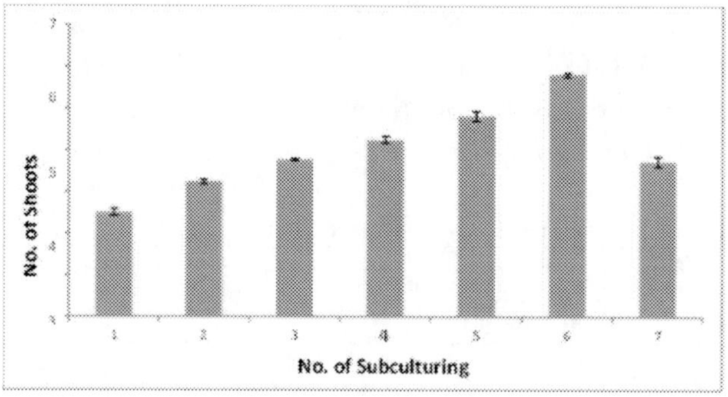

Figure 5. Effect of sub-culturing on the number of shoots per explants.

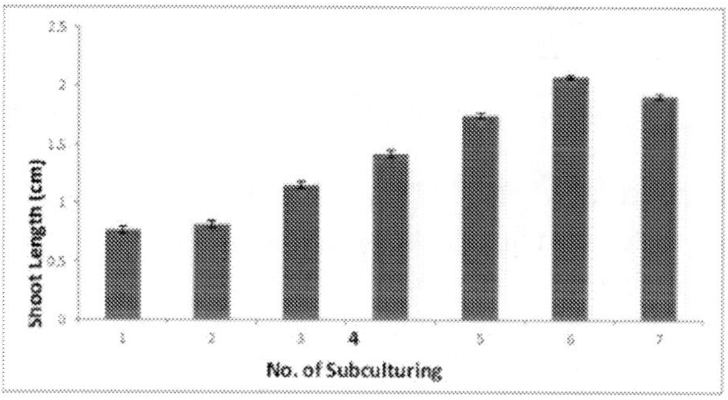

Figure 6. Effect of subculturing on the shoot length per explant.

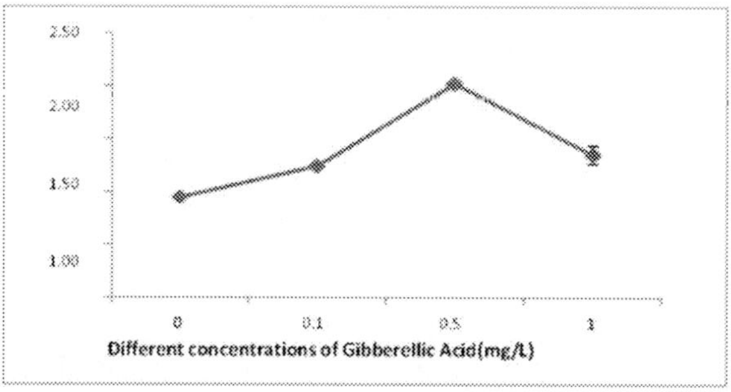

Figure 7. Effect of gibberellic acid on the shoot length per explants.

Table 1. Effect of different media composition (MS, WPM, B5 & MT) and BAP (1 mg/L) on the initiation of bud proliferation and multiple shoot proliferation from nodal explants of *ex-vitro* grown seedlings of C. *macroptera*

Data as (Mean ± S.D.) was recorded after four weeks of culture

Treatment BAP(mg/L)	% of Explants Response (Bud Break)	Mean no. of Shoots obtained per explants Mean ± SE	Mean shootlength(cm) Mean ± SE
MS (Control)	100.00	1.00 ± 0.00	0.56 ± 0.04
MS+1BAP	100.00	3.80 ± 0.16	1.00 ± 0.08
WPM (Control)	100.00	1.00 ± 0.00	0.43 ± 0.05
WPM+ 1BAP	100.00	1.60 ± 0.16	0.73 ± 0.04
B5(Control)	80.00	0.80 ± 1.10	0.83 ± 0.04
B5+ 1BAP	100.00	1.23 ± 0.20	1.23 ± 0.20
MT (Control)	100.00	1.00 ± 0.00	0.56 ± 0.05
MT+ 1BAP	100.00	2.66 ± 0.09	0.90 ± 0.08

Table 2. Effect of sub-culturing on *in vitro* shoot proliferation of *C. macroptera* in MS + 1.0 mg/L BAP. Data as (Mean ± SE)was recorded after four weeks of culture

Subculture Stage	Shoots per explants (Mean ± SE)	Maximum shoot length (cm) (Mean ± SE)
1	2.50 ± 0.08	0.76 ± 0.02
2	3.23 ± 0.05	0.81 ± 0.02
3	3.77 ± 0.02	1.15 ± 0.02
4	4.25 ± 0.08	1.42 ± 0.03
5	4.81 ± 0.12	1.74 ± 0.02
6	5.81 ± 0.04	2.07 ± 0.01
7	3.73 ± 0.12	1.91 ± 0.02

Conclusion

The present study focusses to conserve and utilize various tissue culture techniques for preserving the medicinally important plant C. macroptera. This involves standardization of efficient tissue culture techniques such as micropropagation using nodal explants from in vitro grown seedlings. The work involved the efficacy of different types of Plant Tissue Culture medium on percentage seed germination, combinatorial effect on different concentration of phytohormones and its combinations/adjuvants that were used for the initiation of cultures and shoot proliferation. The results hints at the possibility of single shoot subculturing with subsequent detailed future experiments.

References

Bhattacharya, S. C., & Dutta, S. (1956). Classification of citrus fruits of Assam. Indian Council of Agricultural Research. *ICAR Sci Monog* 20: 1–110

Bitters, W. P. and Murashige, T., 1967, A place for tissue culture in citrus research. *Calif. Citrograph* 52: 226-228.

Bunn, E., Turner, S., Panaia, M., Dixon, K. W., 2007, The contribution of in vitro technologyand cryogenic storage to conservation of indigenous plants. *Australian Journal of Botany* 55(3):345-355.

Button, J. and Kochba, J., 1977, *Applied and Fundamental Aspects of Plant Cell, Tissue and Organ Culture*, (Reinert, J. and Bajaj; Y. P. S. eds.) SpringercVerlag, Berlin, pp. 70-92.

Chen, D., Guo, B., Hexige, S., Zhang, T., Shen, D., and Ming, F. 2007. SQUAlike genes in the orchid Phalaenopsis are expressed in both vegetative and reproductive tissues. *Planta* 226: 369–380.

Coates, D. & Dixon, K. 2007, Current perspectives in plant conservation biology. *Australian Journal of Botany*, 55(3):187–193.

De Almeida, W. A. B., De, F., Mendes, B. M. J., Rodriguez, A. P. M., 2006, Histological characterization of in vitro adventitious organogenesis in *Citrus sinensis*. *Biologia Plantarum* 50:321–325.

Debergh, P., Roggemans, J., Standaert-De Metsenaere, R., 1990, Tissue culture in relation to ornamental plants. *Developments in Crop Science* 19:161–189.

Engelmann, F., 1991, In vitro conservation of tropical plant germplasm - a review. Euphytica. 57: 227-243.

Hamza, A., Gtari, M., & Mohamed, N. 2013, Micropropagation of Helianthemum lippii L. var Sessiliforuim (Cistaceae) an important pastoral plant of North African arid areas. *African Journal of Biotechnology*, 12(46):6468-6473.

Haripyaree, A., Guneshwor, K., Sunitibala, H., Damayanti, M., 2011, In vitro propagation of Citrus megaloxycarpa. *Environmental and Experimental Biology* 9:129–132.

Hussey, G. 1977. In vitro propagation of gladiolus by precocious axillary shoot formation. *Scientia Hort.*, 6:287-289.

Jona, R., Menini, U. G., 1987, Tissue culture of selected tropical fruit plants. *FAO Plant Production And Protection*, 79-128.

Laskar, M. A., Hynniewta, M., Rao, C.S., 2009, In vitro propagation of Citrus indica Tanaka - An endangered progenitor species. *Indian Journal of Biotechnology* 8:311–316.

Ling, J. T., Iwamasa, M., 1997, Plant regeneration from embryogenic calli of six Citrus related genera. *Plant Cell, Tissue and Organ Culture* 49:145–148.

Malik, S. K., Chaudhury, R., Dhariwal, O. P., Kalia, R. K., 2006, Collection and characterization of Citrus indica Tanaka and C. macroptera Montr: Wild endangered species of northeastern India. *Genetic Resources and Crop Evolution.* 53:1485–1493.

Miah, M. N., Islam, S., Hadiuzzaman, S., 2008, An improved protocol for multiple shoot regeneration from seedlings and mature explants of Citrus macroptera Mont. *Plant Tissue Culture and Biotechnology* 18(1): 17-24.

Normah, M. N., Hamidah, S., Ghani, F. D., 1997. Micropropagation of Citrus halimii - An endangered species of South-east Asia. *Plant Cell, Tissue and Organ Culture* 50:225–227.

Parawira, W., & Khosa, E. M. 2009. Biotechnology research, development, applications and management in Zimbabwe: Review. *Scientific Research and Eassay*, 4(9).825-841.

Ruchi V., S. Shukla., 2018, A significant footprint of *In-vitro* micropropagation on growing opportunities of endangered citrus species in India, International Conference On Recent Trends in Agriculture, Food Science, Forestry, Horticulture, Aquaculture, Animal Sciences, Biodiversity, *Ecological Sciences and Climate Change*, 37-41.

Shamsul H. Prodhan, Md. Nazmul Hasan, Hammadul Hoque, S. S. Alam, Mohammed Raqibul Hasan, Anup Gupta, Most. Ummay Salma Khatun, Afroza Parvin, Ziaul Faruque Joy., 2016, Development of an efficient in vitro regeneration system for endangered wild orange citrus Chrysocarpa L., *International Journal of Sciences*: Basic and Applied Research, 30(5): 187- 196.

Sharma, B. D., Hore, D. K., & Gupta, S. G. (2004). Genetic resources of Citrus of north-eastern India and their potential use. *Genetic Resources and Crop Evolution, 51*(4), 411-418.

Swingle, W. T., Reece, P. C., Reuther, W., Webber, H. J., & Batchelor, L. D. (1967). The citrus industry. *Revised, 1.*

Tanaka, T. (1954). Species problem in citrus. *Japanese Society for Promotion of Science, Tokyo.* p.152.

Chapter 12

Optimization of Microwave Sterilization as an Effective Method of Fresh Produce Preservation

Susmita Mukherjee, Riya Chakraborty, Ankush Ghosh, Ritasree Deb Roy and Sonali Paul*

Department of Biotechnology,
University of Engineering and Management,
University Area, Kolkata, West Bengal, India

Abstract

Fruits and vegetables are a natural source of health-enhancing nutrients. These phytonutrients show a high level of antioxidant and anti-inflammatory activities that can ensure a healthy human system. The fungal pathogen *Botrytis cinerea* affects fruits in the field as well as post-harvest during storage, transportation, and marketing, leading to significant economic losses. The use of microwave is gaining popularity as it ensures product quality for processed fruits and vegetables. In the present work, we have observed that an optimum microwave exposure leads to enhancement of the shelf life of fruits and vegetables by delaying the mold infection while retaining the number of essential phytonutrients. Thus it can be concluded that an optimum microwave exposure before storage can increase the shelf life and can be a useful method of preservation of fresh produce.

---

* Corresponding Author's Email: sonali.paul@uem.edu.in.

In: Proceedings of BIOSPECTRUM
Editors: Sanket J. Joshi, Susmita Mukherjee and Moupriya Nag
ISBN: 978-1-68507-985-7

Keywords: food preservation, grey mold, microwave, phytonutrients, shelf life

Introduction

Microwave sterilization is a thermal process (US FDA, 2000). It delivers microwave radiation to the food package under controlled pressure and temperature to achieve the inactivation of pathogenic bacteria. Currently, processed foods are heat-treated for sterilization. Microwaves are in contact with polar water molecules and charged ions. The friction resulting from molecules rapidly aligning in electromagnetic fields generates heat directly within the food that sharply reduces the thermal processing time. The shorter time required alongwith more uniform heating during microwave processing help in better retainment of the sensory attributes of foods [1, 2]. Compared with conventional food processing, microwave sterilization offers a potentially better nutritionally preserved food product [3]. Microwave sterilization can effectively retain the nutrients in fruit juice [4]. Microwave sterilization of rice decreased the spore number of *Aspergillus niger* [5, 6]. Microwave sterilization had no significant effect on the colour of fruit and vegetable products, could retain the nutrients such as phenols, flavonoids, and vitamin C, had a strong inhibitory effect on polyphenol oxidase, and peroxidase. A high-power microwave would cause water loss, thus damaging the organizational structure of fruits and vegetables [7]. The lethal effect of microwaves on different spores of *Bacillus* sp. was studied. Due to their efficient and uniform heating effect, microwave sterilization is the future of bioprocessing techniques [8].

Methods

Samples Taken

Kiwi, apple, bell pepper was subjected to microwave exposure (800W, 2450MHz) for different times of exposure –30 secs, 60 secs, 90 secs, and control with no microwave exposure.

Microbiological Examination

The molds from the surface of the samples were taken on a microscopic slide, stained with lactophenol cotton blue, and observed under the microscope [6].

Estimation of Total Carbohydrate

The samples (100mg) were acid digested and then neutralized with sodium carbonate, centrifuged, supernatant collected. To it anthrone was added, boiled, cooled and the absorbance was recorded at 630nm [9].

Estimation of Ascorbic Acid

The sample extract was prepared by grinding the samples (0.5g) in 4% oxalic acid, centrifuging, collecting the supernatant, adding oxalic acid, and titrating against the dye which is 2, 6 dichlorophenol indophenol. The appearance of pink colour signals the end of the reaction. The amount of dye consumed gives the amount of ascorbic acid [9].

Estimation of Vitamin E

Sample extract was prepared by homogenizing 0.5g of the sample, adding 0.1N sulphuric acid, stoppered, and being allowed to stand overnight. It was then filtered. To the filtrate, ethanol was added, centrifuged, supernatant collected. To the supernatant, xylene was added, centrifuged and the xylene layer collected. To the xylene, a layer of 2,2' – dipyridyl was added, and absorbance was measured at 460nm. Then ferric chloride solution was added and absorbance measured at 520nm against standard DL-α-tocopherol [9].

Estimation of Flavonoids

The samples were boiled in water-ethanol solution 60% (v/v) for 60 min. It was then cooled and filtered. To the filtrate (0.08g / ml), methanol and 2%

$AlCl_3$ were added and incubated for 25 min. The absorbance was recorded at 430 nm [10].

Estimation of Total Phenol

The sample extract was prepared by grinding the samples (10g) in 80% ethanol and collecting the supernatant. The supernatant was air-dried followed by dissolution in distilled water. To small aliquots, Folin Ciocalteau reagent and sodium carbonate was added. The tubes were placed in boiling water, cooled and the absorbance was recorded at 650nm. The total phenol concentrations were calculated by graphical plotting against a standard curve prepared by using varying concentrations of catechol [9].

Estimation of Total Carotenoids

The sample (10g) was ground in 3% acetone and the supernatant collected for preparation of the sample extract. This was followed by the addition of petroleum ether and distilled water. The carotenoid pigments enter the petroleum ether phase. The absorbance was recorded at 452nm. against standard concentrations of β carotene [9].

Statistical Analysis

All experiments were conducted in triplicate, standard deviation determined, and students t-test at $p < 0.05$ applied to find out the statistical significance.

Results

Physical Observation

All the samples were kept at room temperature and the first appearance of mold was noted in the control samples. It was observed that the first appearance of mold was delayed in microwave exposed samples with a correlation between the time of exposure and the time of the first appearance of mold (Table 1). Molds appeared as small spots which later on completely covered the surface of the sample which looked like large grey areas covered with mycelium.

Microbiological Examination

Observation under the microscope revealed the presence of branched mycelium with conidiophores growing directly from mycelia, the density of which increased from the time the first spot appeared. The conidia were grape-shaped. The causal organism is Botrytis cinerea which forms grey mold.

Estimation of Total Carbohydrate

Carbohydrate is one of the primary nutrients of all of our study samples. There were differences in the total carbohydrate content of the samples with differences in times of microwave exposure (Table 2). The carbohydrate content decreased with an increase in microwave exposure time. The greatest decrease was observed in the sample having 90 secs exposure time and the lowest decrease observed in an exposure time of 30secs.

Estimation of Ascorbic Acid

Ascorbic acid found in fruits and vegetables helps in enhancing the immune response of our body. All the study samples had a substantial amount of ascorbic acid which decreased with microwave exposure as compared with control (Table 3). The greatest decrease was observed in the sample having the highest exposure of 90 secs and the lowest decrease observed in the least exposure time of 30secs.

Estimation of Vitamin E

Vitamin E is one important nutrient that functions as an antioxidant. Kiwi is a fruit that serves as an important source of vitamin E. There was a decrease in the vitamin E content with the time of microwave exposure (Table 4). The control sample contains 4.6 mg / g vitamin E, the sample with 30secs exposure contains 4.5 mg / g vitamin E, the sample with 60 secs exposure contains 4.2 mg / g vitamin E, and the sample with 90 secs exposure contains 3.7 mg / g vitamin E. Thus the sample with 30 secs exposure had 2.17% decrease, the sample with 40 secs exposure had 8.7% decrease and the sample with 90 secs exposure had 19.6% decrease as compared with control.

Estimation of Flavonoids

Flavonoids are a group of plant metabolites having antioxidant effects. Apple and bell pepper has a high content of flavonoids which decreased with different periods of microwave exposure as compared with control (Table 5). The greatest decrease was observed in the sample having the highest exposure of 90 secs and the lowest decrease observed in the least exposure time of 30secs.

Estimation of Total Phenol

Polyphenols are plant secondary metabolites abundantly occurring in different fruits like apples which impart different health benefits. There was a decrease in the polyphenol content with the time of microwave exposure (Table 6). The control sample contains 7.1 mg of GAE / g total phenol, the sample with 30secs exposure contains 6.9 mg of GAE / g total phenol, the sample with 60 secs exposure contains 6.8 mg of GAE / g total phenol, and the sample with 90 secs exposure contains 6.2 mg of GAE / g total phenol. Thus the sample with 30 secs exposure had a 2.82% decrease, the sample with 40 secs exposure had a 4.23% decrease, and the sample with 90 secs exposure had a 12.68% decrease as compared to the control sample.

Estimation of Total Carotenoids

Carotenoids are important phytonutrients that function as an antioxidant by deactivating free radicals within the human body. There was a decrease in the total carotenoid content with the time of microwave exposure (Table 7). The control sample contains 5.95 mg / g total carotenoids, the sample with 30secs exposure contains 5.8 mg / g total carotenoids, the sample with 60 secs exposure contains 5.6 mg / g total carotenoids, and the sample with 90 secs exposure contains 5.1 mg / g total carotenoids. Thus the sample with 30 secs exposure had a 2.52% decrease, the sample with 40 secs exposure had a 5.88% decrease, and the sample with 90 secs exposure had a 14.29% decrease as compared with the control sample.

Discussion

Physical and Microbiological Observation

Grey mold caused by Botrytis cinerea is one of the most prevalent post-harvest diseases [11]. Phytochemicals including phenolics, flavonoids, and carotenoids play an important role in decreasing the risk of chronic diseases [12]. Higher mold infections in fruits and vegetables occur in tropical countries like India which is characterized by hot and humid environmental conditions [13]. This leads to the requirement of devising strategies for the preservation of fruits and vegetables to minimize this economic loss. Microwaves destroy mold spores in a very short exposure time [14]. In our work too, we have observed delayed grey mold infection with increased microwave exposure of the study samples.

Table 1. The appearance of mold (days)

	Kiwi	Apple	BellPepper
Control	7	15	5
30 secs	9	20	7
60 secs	15	31	10
90 secs	16	33	12

Total Carbohydrate Content in Food

According to the U. S. Department of Agriculture (USDA) Dietary Guidelines 2005, the best carbohydrates are those that contain a lot of fibre, like vegetables, fruits, and whole grains. It has been found that there is greater retention of micronutrients with microwave cooking in comparison to other cooking methods due to reduced cooking time. A vast majority of nutrients are lost by getting leached into the cooking water. The fact that microwaves allow for "water-free" cooking is highly beneficial [15]. In our work we have found some amount of decrease in carbohydrate content with greater microwave exposure leading to greater decrease as the destruction of nutrients depends on the exposure time with less exposure time leading to less destruction of nutrients [16].

Table 2. Total carbohydrate content (mg/ml)

	Kiwi	Apple	BellPepper
Control	11.9	10.2	6
30 secs	11.5	9.9	5.85
60 secs	11.4	9.6	5.7
90 secs	10.6	9.4	5.0

The Ascorbic Acid Content in Food

Vitamin C, also known as ascorbic acid, acts as an antioxidant, helping to provide protection against cellular damage caused by free radicals [17]. Our study samples showed the presence of ascorbic acid which decreased in the microwave treated samples as compared to the control, with a greater decrease observed with greater microwave exposure. This may be because ascorbic acid undergoes thermal degradation with greater degradation observed on higher heat exposure [18].

Table 3. Total ascorbic acid content (mg/g)

	Kiwi	Apple	BellPepper
Control	4.1	1.2	1.6
30 secs	4.0	1.0	1.6
60 secs	3.5	0.8	1.2
90 secs	3.0	0.5	0.9

Vitamin E Content in Food

Vitamin E is a fat-soluble vitamin existing in several forms with alpha-tocopherol being the only one used by the human body. Its main role is to act as an antioxidant, scavenging loose electrons—so-called "free radicals"—that can damage cells. It also enhances immune function and prevents artery clotting [19]. We observed increased degradation of vitamin E content with increased microwave exposure. This is since vitamin E is sensitive to microwave exposure with the amount of decrease correlated to the temperature and time of exposure [20].

Table 4. Total vitamin E content (mg/g)

	Kiwi
Control	4.6
30 secs	4.5
60 secs	4.2
90 secs	3.7

Flavonoids Content in Food

Flavonoids exists in the form of glycosides. During the process of digestion, most flavonoid glycosides pass into the small intestine and are decomposed by the intestinal microflora in the colon. Some of them are absorbed from the small intestine and function to prevent the oxidation of low-density lipoproteins [21]. Microwave heating is a comparative better method for retention of nutrients as compared to other methods of food processing as the addition of water is not a prerequisite in this method [22]. In our work, we have observed that less heat exposure leads to lesser loss of flavonoids in food.

Table 5. Total flavonoid content (mg/g)

	Apple	Bell Pepper
Control	1.2	4.1
30 secs	1.2	4
60 secs	0.9	3.8
90 secs	0.5	3.1

Phenol Content in Food

Phenols are one of the major groups of dietary components appearing in vegetable foods. They are a wide group including more than 8,000 different compounds that are considered secondary plant metabolites. The bioactivity of phenolics is attributed to their ability to chelate metals, inhibit lipoxygenase and scavenge free radicals thereby inhibiting diseases like atherosclerosis and cancer [23]. Our work demonstrates less degradation of phenol compounds with less duration of microwave exposure to fruits and vegetables [24].

Table 6. Total phenol content (mg of GAE/g)

	Apple
Control	7.1
30 secs	6.9
60 secs	6.8
90 secs	6.2

Carotenoid Content in Food

Dietary carotenoids including beta-carotene, lycopene, lutein, and zeaxanthin are thought to decrease the risk of certain cancers and eye diseases. In part, the beneficial effects of carotenoids are thought to be due to their role as antioxidants [25]. In our work the carotenoid content of bell pepper decreased with microwave exposure as carotenoids are temperature sensitive with higher loss observed during greater microwave exposure [26].

Table 7. Total carotenoid content (mg/g)

	Bell Pepper
Control	5.95
30 secs	5.8
60 secs	5.6
90 secs	5.1

Conclusion

From the above discussion, it is clear that grey mold caused by Botrytis cinerea leads to post-harvest economic loss during fruit and vegetable storage. From the results of the present work, it can be concluded that an optimum microwave exposure of 800W for 30 secs helps in reducing the incidence of grey mold along with retention of vital phytonutrients. Thus this method provides for a simple, economically feasible, efficient method of food preservation.

References

[1] Guan, D., Gray, P., Kang, DH, Tang, J., Shafer, B., Ito, K., Younce, F., and Yang, T.C.S. 2003. Microbiological validation of microwave-circulated water combination heating technology by inoculated pack studies, *J. Food Sci.,* 68(4):1428-1432.

[2] Tang, Z., Mikhaylenko, G., Liu, F., Mah, JH., Tang, J., Pandit, R., Younce, F., 2008. Microwave sterilization of sliced beef in gravy in 7 oz trays. *J. Food Engineering.* 89(4):375-383.

[3] Guan, D., Plotka, V. C. F., Clark, S., and Tang J. 2002. Sensory evaluation of microwave treated macaroni and cheese. *J. Food Processing and Preservation,* 26:307-322.

[4] X. L. Yu, R. C. Shi, L. Wang. 2016. Effect of microwave treatment on the quality of Carica Papaya L. puree. *Food Science and Technology*, 5: 64-70.

[5] Y. Y. Xu, K. J. Li, Y. Qiu. 2014. Process optimization of microwave sterilization for aspergillus niger from rice. *Science and Technology of Food Industry*, 35(22): 245-248,253.

[6] Zhou X, Li Y. 2015. Techniques for oral microbiology. *Atlas of Oral microbiology*. Chapter 2. 15-40.

[7] Zhu X, Yang Y, Duan Z. 2018. Research progress on the effect of microwave sterilization on agricultural products quality. *Earth and Environmental Science* 113(1):012096.

[8] Orsat V., Raghavan G.S.V. 2007. Microwaves in postharvest applications with fresh fruits and vegetables. *Fresh Produce*. 1(1): 16-22.

[9] Sadasivam S. and Manickam A. 2008. *Biochemical Methods*. New Age International (P) Ltd., Publishers, New Delhi, India.

[10] Da Silva L. A. L., Pezzini B.R., Soares L. 2015. Spectrophotometric determination of the total flavonoid content in *Ocimum basilicum* L. (Lamiaceae) leaves. *Pharmacogn. Mag*. 11: 96–101.

[11] Bautista-Baños, S., Long, P. G., and Ganesh, S. 1997. Curing of kiwifruit for control of postharvest infection by Botrytis cinerea. *Postharvest Biol. Technol*. 12:137-145.

[12] Boyer J, Liu R. H. 2004. Apple phytochemicals and their health benefits. *Nutr. J.* 3:5.

[13] Zoffoli J.P. and Latorre BA. 2011. 9-Table Grape (Vitis vinifera L.). *Postharvest Biology and Technology of Tropical and Subtropical Fruits*. Woodhead Publishing Series 179-212.

[14] Michaelson SM. 1974 Effects of exposure to microwaves: problems and perspectives. *Environ Health Perspect*. 8:133-55.

[15] Kimura M, Itokawa Y. 1990. Cooking losses of minerals in food and its nutritional significance. *J Nutr Sci Vitaminol*. 36 (1): S25-32.

[16] Cross GA, Fung DY. 1982. The effect of microwaves on the nutrient value of foods. *Crit. Rev. Food Sci Nutr.* 16(4):355-81.

[17] Machlin LJ, Bendich A. Free radical tissue damage: protective role of antioxidant nutrients. *The Faseb Journal*. 1987. 1 (6) : 441-445.

[18] Vernin G, Chakib S, Rogacheva SM, Obretenov TD, Parkanyi C. Thermal decomposition of ascorbic acid. *Carbohydrate Research*. 1997. 305 (1): 1-15.

[19] Stampfer MJ, Hennekens CH, Manson JE, Colditz GA, Rosner B, Willett WC. 1993. Vitamin E consumption and the risk of coronary disease in women. *N. Engl. J. Med.* 328: 1444-1449.

[20] Sieber R, Eberhard P, Fuchs D, Gallmann PU, Strahm W. 1996. Effect of microwave heating on vitamins A, E, B1, B2, and B6 in milk. *Journal of Dairy Research*. 63: 169-172.

[21] Lean ME, Noroozi M, Kelly I, Bruns J, Talwar D, Sattar N, Crozier A. 1999. Dietary flavonols protect diabetic human lymphocytes against oxidative damage to DNA. *Diabetes* 48: 176-181.

[22] Ioku K, Aoyama Y, Tokuno A, Terao J, Nakatani N, Takei Y. 2001. Various Cooking Methods and the Flavonoid Content in Onion. *J Nutr Sci Vitaminol*, 47: 78-83.

[23] Martínez-Valverde I, Periago MJ, Ros G. Nutritional importance of phenolic compounds in the diet. 2000. *Arch Latinoam Nutr*. 50(1):5-18.

[24] Cerretani L, Bendini A, Rodriguez-Estrada MT, Vittadini E, Chiavaro E. 2009. Microwave heating of different commercial categories of olive oil: Part I. Effect on chemical oxidative stability indices and phenolic compounds. *Food Chemistry*. 115 (4):1381-1388.

[25] K. J. Li, Y. Y. Xu, H. H. Guan, 2015. Microwave sterilization of aspergillus flavus in rice. *Journal of Food Safety and Quality Inspection*, 9: 3684-3691.

[26] Fratianni A, Cinquanta L, Panfili G. 2010. Degradation of carotenoids in orange juice during microwave heating. LWT - *Food Science and Technology*. 43 (6): 867-871.

Chapter 13

Effect of Incorporation of Unsaponifiable Matter Isolated from Rice Bran Oil Fatty Acid Distillate on Antioxidant Activity and Storage Stability of Edible Oils

Sudhasri Sahu*, Swarnali Mitra, Minakshi Ghosh and D. K. Bhattacharyya

School of Community Science and Technology,
Indian Institute of Engineering Science and Technology, Shibpur, India

Abstract

The antioxidant activity and storage stability of two edible vegetable oils, such as Sunflower oil and Soybean oil incorporated with unsaponifiable matter (γ-oryzanol, total tocopherol, total phytosterol and squalene) isolated from rice bran oil fatty acid distillate (RBOFAD) were investigated. The physicochemical characteristics and antioxidant activities of the individual edible oils as such and containing the unsaponifiable matter (UM) were evaluated based on free fatty acid (FFA), peroxide value (PV) and *p*-anisidine value (*p*AV), DPPH Scavenging activity and γ-oryzanol content. This investigation showed that FFA, PV, PAV, DPPH Scavenging activity and γ-oryzanol content significantly changed during 60 days of storage under ambient conditions and darkness at room temperature. The unsaponifiable matter of initial rice bran fatty acid distillate contained γ-oryzanol (3.27 gm/100gm), total tocopherol (10.93 mg/100gm), total phytosterol (21.81 gm/100gm), squalene (1.15 gm/100gm) and fatty alcohol (73.34 gm/100gm). The

* Corresponding Author's Email: sudhasri.sahu@gmail.com.

In: Proceedings of BIOSPECTRUM
Editors: Sanket J. Joshi, Susmita Mukherjee and Moupriya Nag
ISBN: 978-1-68507-985-7

high amount of fatty alcohol was separated from the unsaponifiable matter of rice bran oil fatty acid distillate by the alcohol extraction process at room temperature. The antioxidant capacity of the unsaponifiable matter of RBOFAD (UMRBOFAD) reduced the extent of oxidation of the two vegetable oils when blended with isolated unsaponifiable matter from RBOFAD.

Keywords: rice bran oil fatty acid distillate (RBOFAD), unsaponifiable matter of RBOFAD (UMRBOFAD), storage stability, antioxidant activity

Introduction

Rice bran oil is a significant by-product of oil, and it is regarded as an extremely valuable edible oil from nutrition and health aspects. The oil is characterized by its unique fatty acid composition having in nearly equal proportion of saturated, monounsaturated, polyunsaturated fatty acids along with the unique combination of unsaponifiable matters (Van Hoed et al. 2006; Ghosh 2007). Rice bran oil is refined by the physical refining process after degumming, bleaching and dewaxing in order to retain the maximum extent of unsaponifiable matters for their bioactivity. The physical refining process produces distillate products known as rice bran oil fatty acid distillate which comprises, besides fats also some of the unsaponifiable matters of rice bran oil having bioactive properties. These unsaponifiable matters are presently composed of oryzanol, tocopherols, squalene, sterols, etc. (Nandi and Bhattacharyya 2013; Rodrigues et al. 2014; Pal and Pratap 2017). γ-Oryzanol, a natural antioxidant in rice bran oil, is associated with decreasing plasma cholesterol, decreasing platelet aggregation and lowering cholesterol absorption. γ-oryzanol was used as a potential additive in many food sectors, cosmetics and pharmaceutical industries (Sohail et al. 2017; Sahu et al. 2018). Tocopherols and tocotrienols (α, β, γ and δ) are also important phytochemicals in rice bran oil with antioxidant activity, and they have potential health benefits. These phytochemicals prevent lipid peroxidation for the storage stability of vegetable oils. Tina A. Gomes reported the isolation of tocols rich fraction from the deodorized distillate of rice bran oil (Yang, Chen, and Yang 2018; Chen and Bergman 2005; Gomes et al. 2015). The phytosterols such as campesterol, β-sitosterol, stigmasterol and isofucosterol of rice bran oil inhibit the absorption of cholesterol in the small intestine of the human body, and these constituents are more useful substances for utilization of functional food for human health (Nijjar et al. 2010; Gupta et al. 2011; Yamamoto and Har 2012). Squalene, a trace component, is generally found in marine animal oils, and it prevents many diseases like cancer and cardiovascular diseases, and it is beneficial for the skin. Rice bran oil deodorized distillate contains about 8% squalene as unsaponifiable components. Recently squalene is also getting attention as a food nutrient in various food sectors (Bhilwade et al. 2010; Smith 2000). The fatty alcohols are mainly derived from different vegetable oils like rice bran oil sunflower oil used as raw materials for making oleogels that are more important for the bakery industry, and it is utilized in various household products surfactants (Troni et al. 2013; Sahu, Ghosh, and Bhattacharyya 2020). So, the isolation of unsaponifiable matters from rice bran oil fatty acid distillate is of interest and incorporation in food products for more nutritional benefits and stability from oxidation. The natural antioxidants were added in various vegetable oils like sunflower oil, groundnut oil, mustard oil, etc., for increasing their oxidative stability and their shelf life. These supplemented antioxidants were more effective for the protection of oils from lipid oxidation by quenching the free radical reactions (Shiela, Sreerama, and Gopala Krishna 2004; van Aardt et al. 2004).

In India, the various major vegetable oils such as sunflower, soybean, groundnut, mustard are used as cooking oils. These types of cooking oils do not contain valuable health-improving minor components like γ-oryzanol, tocopherol, etc. [16]. So, these types of minor components are important for blending in vegetable oils (sunflower oil, groundnut oil, soybean oil, etc.), and various storage stability of blended oils have been reported. Various previous reports on blended oils have focused on tocopherol components (Li et al. 2013; Chu and Kung 1998).

This investigation attempts to study the antioxidant activity and storage stability of vegetable oils blended with unsaponifiable matter from RBOFAD at room temperature.

Materials and Methods

Materials

The fatty acid distillate of rice bran oil was supplied from Sethia Oils Ltd. (Burdwan, West Bengal, India). Refined vegetable oils such as sunflower oil and soybean oil were purchased from the local market of Kolkata. All samples were stored in amber bottles at room temperature until they could be analyzed. Analytical grade reagents were used for investigation.

Quality Assessment of Fatty Acid Distillate from Rice Bran Oil

Qualities of the fatty acid distillate from rice bran oil were measured through the chemical properties of fatty acid distillate containing the acid value (AOAC 969.17), saponification value (AOAC 920.160), unsaponifiable matter (AOAC 972.28) and peroxide value (AOAC 965.33) following standard AOAC methods (Sahu et al. 2018).

Isolation of Unsaponifiable Matter from RBOFAD

The unsaponifiable matter was isolated from RBOFAD following the standard AOAC method (AOAC 972.28) (Sahu et al. 2018). A high amount of fatty alcohol was present in the unsaponifiable matter, so fatty alcohol was separated from unsaponifiable matter through the alcohol extraction process at room temperature because fatty alcohols are insoluble in alcohol (ethanol).

Determination of Minor Components of Unsaponifiable Matter

γ-Oryzanol content (%) of UMRBOFAD and vegetable oils were determined from spectrophotometer (JASCO V630 UV Vis Mississippi, USA) absorption measurements at the wavelength of maximum absorption near 314nm (Sahu et al. 2018). γ- Oryzanol content (g) percentage was calculated according to this formula shown below, using specific extinction of 358.9 for 1% solution of 100% pure γ-oryzanol.

$$\gamma - Oryzanol content,(g)\ \% = \frac{OD\ Value\ of\ hexane\ solution}{Weight\ of\ oil\ (g)\ X\ 10}\ X\ \frac{100}{358.9}$$

Total tocopherol analysis of RBOFAD was done by HPLC system (Agilent Technologies) equipped with Agilent Zorbax RX – SIL (5μm, 4.6 ×

250mm) column and mobile phase 3.85% Tetrahydrofuran in n-Heptane and Fluorescence detection (Excitation -295nm and Emission-330nm) was used for analysis (Sahu et al. 2018).

After isolation of unsaponifiable matter, the UM was analyzed by Agilent Gas Chromatography (GC) with a capillary column and Flame Ionization Detector (FID) at 350°C for determination of phytosterol, squalene and fatty alcohol (Sahu et al. 2018). The total Phytosterol, squalene and fatty alcohol content were identified by comparing the retention time of phytosterol, squalene and fatty alcohol in these samples with the internal standard for quantification.

Preparation of Vegetable Oils Blended with the Unsaponifiable Matter of RBOFAD

Vegetable oils such as soybean oil (SBO) and sunflower oil (SFO) were blended by mixing with fatty alcohol-free unsaponifiable matter from RBOFAD in varying percentages (0.1 wt%, 0.2 wt% and 0.5 wt%) at 55°C-60°C by stirring on a magnetic stirrer (REMI Magnetic stirrer with hot plate, 5 MLH Plus) for 1 hour. The unsaponifiable matter blended oils thus obtained were stored at room temperature for 60 days.

Analytical Methods

Free fatty acid (FFA as oleic; Wt.%) and peroxide value (PV; meq of O2 per kg of oil) were measured following AOAC methods (AOAC 2005) while *p*-anisidine values (*p*AV) of vegetable oils following AOCS official method Cd 18-90 (Al-Kharousi et al. 2016; Sahu et al. 2018) as such and after blending with the unsaponifiable matter were determined to ascertain oxidative stability that refers to the formation of hydroperoxide and its decomposition products in the chain of unsaturated fatty acids of edible oils.

Free Radical Scavenging Activity (DPPH Assay)

The DPPH (1,1- diphenyl-2-picrylhydrazyl) was measured for the determination of radical scavenging power of minor components enriched vegetable oil samples (0 days to 60 days). These samples were diluted to varying concentrations with isopropanol, and these diluted samples were mixed with 1 ml DPPH free radical solution in a dark place. After 30 min incubation period, this mixture was measured at 540 nm using a UV spectrophotometer (JASCO V630 UV Vis Mississippi, USA) (Lee et al. 2015). The radical scavenging percentage (%) was calculated by the following equation:

$$Free\ radical\ scavenging\ activity\ (I\%) = \frac{A_C - A_S}{A_C} X\,100$$

Where A_C = Absorbance of control; A_S = Absorbance of the sample at 517 nm.

γ -Oryzanol Content Determination

γ-Oryzanol content (%) of oils were determined from spectrophotometer absorption measurements at the maximum absorption wavelength at 314nm as described earlier.

Statistical Analysis

Statistical analysis was done by Tukey test ($p < 0.05$) for intergroup comparison of parametric data. ANOVA One-way analysis was performed in this method by using Origin 8 software.

Results and Discussion

Chemical Characteristics of the Original Fatty Acid Distillate of Rice Bran Oil

The chemical characteristics of rice bran oil fatty acid distillate had acid value (163.66 ± 0.57 mg of KOH per g of oil), saponification value (181.38 ± 5.66 mg of KOH per gm of oil), peroxide value (8.33 ± 2.88 meq of O2 per kg of oil) and unsaponifiable matter (11.30 ± 0.26 Wt.%). These are essential analytical parameters for quality measurements of vegetable oils.

Isolation of Unsaponifiable Matter from RBOFAD

Unsaponifiable matter (11.30 Wt.%) was isolated from rice bran oil fatty acid distillate. Various minor components such as γ-oryzanol, tocopherol, phytosterol, squalene and fatty alcohol are present in this unsaponifiable matter, as shown in Table 1. High amount of fatty alcohols present in unsaponifiable matter, needs separation from unsaponifiable matter for quality improvement of unsaponifiable matter.

Storage Behaviour of Individual Edible Oils and When Blended with Unsaponifiable Matter

The free fatty acid (FFA as oleic), peroxide value (PV) and *p*-anisidine value (*p*AV) of the individual edible oils as such and containing the unsaponifiable matter are shown in Table 2, 3 and 4. The FFA content of UM blended oils slightly decreased during 60 days of storage than the initial FFA content of base oils (Table 2). This result did not indicate rancidity as there was no off flavour normally formed from the decomposition of hydroperoxide in both the individual oils, which was regarded as base oils and also for the blended oils when contained UM due to the presence of antioxidants such as tocopherol. Tocopherol is an important natural antioxidant that defends the oxidation of vegetable oil during storage periods (Al-Kharousi et al. 2016). The FFA content of base oils (sunflower oil and soybean oil) was higher than UM mixed oils (SFO+ UMRBOFAD; SBO+UMRBOFAD) during storage time. PV and *p*-anisidine value of UM blended oils slightly changed compared to the individual base edible oils (sunflower oil, soybean oil) during storage periods at room temperature (Table 3 and Table 4).

Table 1. Compositions of unsaponifiable matter (UM) from rice bran oil fatty acid distillate

Unsaponifiable Matter Components	
γ-oryzanol (gm/100gm)	3.27 ± 0.03
Total Tocopherol (mg/100gm)	10.93 ± 0.01
Total Phytosterol (gm/100gm)	21.81 ± 0.01
Squalene (gm/100gm)	1.15 ± 0.01
Total fatty alcohol (gm/100gm)	73.34 ± 0.01

Each value is an average of three determinations, mean ± SD

Table 2. FFA content of individual vegetable oils with and without unsaponifiable matter during storage period in days

	FFA as oleic (Wt.%)			
Edible Oils + UMRBOFAD (Wt. %)	0 Days	15 Days	30 Days	60 Days
SFO	0.216 ± 0.005	0.280 ± 0.010^{b}	0.336 ± 0.005^{b}	0.280 ± 0.005^{b}
SFO+UM (Wt. 0.10%)	0.221 ± 0.100	0.290 ± 0.005^{b}	0.276 ± 0.003^{a}	0.271 ± 0.003^{a}
SFO+UM (Wt.0.20%)	0.165 ± 0.008	0.248 ± 0.008^{a}	0.335 ± 0.050^{a}	0.249 ± 0.050^{a}
SFO+ UM (Wt.0.50%)	0.358 ± 0.080	0.263 ± 0.020^{ns}	0.399 ± 0.075^{ns}	0.257 ± 0.000^{ns}
SBO	0.363 ± 0.077	0.135 ± 0.005^{a}	0.288 ± 0.252^{ns}	0.272 ± 0.222^{ns}
SBO+UM (Wt.0.10%)	0.294 ± 0.061	0.144 ± 0.006^{a}	0.173 ± 0.009^{a}	0.160 ± 0.009^{a}
SBO+UM (Wt.0.20%)	0.153 ± 0.006	0.160 ± 0.009^{ns}	0.150 ± 0.008^{ns}	$0.176 \pm .010^{a}$
SBO+UM (Wt.0.50%)	0.168 ± 0.009	0.146 ± 0.005^{a}	0.163 ± 0.006^{ns}	0.144 ± 0.008^{a}

SFO = Sunflower oil, SBO = Soybean oil; UMRBOFAD = Unsaponifiable matter of rice bran oil fatty acid distillate. At $p < 0.05$, significant differences are shown by symbol (a-b). ns = Not significant.

Table 3. PV of base oils (without UM) and blended oils with UM during storage periods

	PV (meq O2/Kg oil)			
Edible Oils + UMRBOFAD (Wt. %)	0 Days	15 Days	30 Days	60 Days
SFO	0.41 ± 0.02	0.54 ± 0.04^{a}	0.89 ± 0.03^{b}	1.45 ± 0.05^{b}
SFO + UM (Wt.% 0.10)	0.55 ± 0.04	0.49 ± 0.04^{ns}	1.31 ± 0.12^{b}	2.93 ± 0.36^{b}
SFO + UM (Wt.% 0.20)	0.79 ± 0.03	1.00 ± 0.06^{a}	1.48 ± 0.12^{b}	3.76 ± 0.06^{b}
SFO + UM (Wt% 0.50)	0.88 ± 0.07	1.10 ± 0.01^{a}	1.72 ± 0.08^{b}	4.10 ± 0.10^{b}
SBO	0.08 ± 0.07	0.21 ± 0.03^{ns}	1.22 ± 0.15^{b}	1.51 ± 0.07^{b}
SBO + UM (Wt.% 0.10)	1.10 ± 0.10	1.52 ± 0.06^{a}	1.71 ± 0.01^{a}	1.51 ± 0.10^{a}
SBO + UM (Wt.% 0.20)	1.08 ± 0.12	1.42 ± 0.11^{a}	1.65 ± 0.08^{a}	1.83 ± 0.13^{a}
SBO + UM (Wt.% 0.500	1.10 ± 0.10	1.23 ± 0.12^{ns}	1.62 ± 0.13^{a}	1.52 ± 0.06^{a}

SFO = Sunflower oil, SBO = Soybean oil; UMRBOFAD = Unsaponifiable matter of rice bran oil fatty acid distillate. At $p < 0.05$, significant differences are shown by symbol (a-b). ns = Not significant.

DPPH Free Radical Scavenging Activity

Antioxidant properties of the individual oils as such and containing UM have been studied through DPPH scavenging activity, as shown in Figure 1. The DPPH radical-scavenging activity of minor components enriched oils were higher (SBO; I% = 27.40, SFO; I% = 26.90) than the individual oils (SBO; I% = 15.24, SFO; I% = 16.87) because the blended edible oils contained high amount of unsaponifiable matter (γ-oryzanol, tocopherols, sterols, squalene) which has more antioxidant activity. γ-Oryzanol, tocopherol are natural antioxidants in rice bran oil (Ghosh 2007). γ-Oryzanol content of blended vegetable oils with UMRBOFAD was higher than individual base oils (sunflower oil and soybean oil) according to Table 5.

Table 4. pAV of individual base oils and blended oils containing UM during storage periods

	*p*AV			
Edible Oils + UMRBOFAD (Wt.%)	0 Days	15 Days	30 Days	60 Days
SFO	3.22 ± 0.20	4.32 ± 0.07^{b}	4.70 ± 0.09^{b}	6.05 ± 0.06^{b}
SFO + UM (Wt.% 0.10)	5.84 ± 0.05	6.10 ± 0.10^{a}	4.26 ± 0.25^{b}	9.00 ± 0.10^{b}
SFO + UM (Wt.% 0.20)	3.88 ± 0.05	4.12 ± 0.08^{a}	4.86 ± 0.08^{b}	9.16 ± 0.11^{b}
SFO + UM (Wt% 0.50)	7.65 ± 0.26	9.25 ± 0.23^{a}	9.53 ± 0.13^{b}	9.31 ± 0.28^{a}
SBO	2.13 ± 0.15	2.30 ± 0.12^{ns}	2.72 ± 0.12^{a}	9.21 ± 0.17^{a}
SBO + UM (Wt.% 0.10)	1.58 ± 0.50	2.11 ± 0.10^{ns}	5.25 ± 0.05^{b}	8.53 ± 0.13^{b}
SBO + UM (Wt.% 0.20)	6.72 ± 0.16	7.15 ± 0.11^{a}	7.16 ± 0.05^{a}	7.63 ± 0.13^{a}
SBO + UM (Wt.% 0.500	7.13 ± 0.0.08	7.94 ± 0.05^{b}	8.55 ± 0.21^{b}	9.32 ± 0.10^{b}

SFO = Sunflower oil, SBO = Soybean oil; UMRBOFAD = Unsaponifiable matter of rice bran oil fatty acid distillate. At $p < 0.05$, significant differences are shown by symbol (a-b). ns = Not significant.

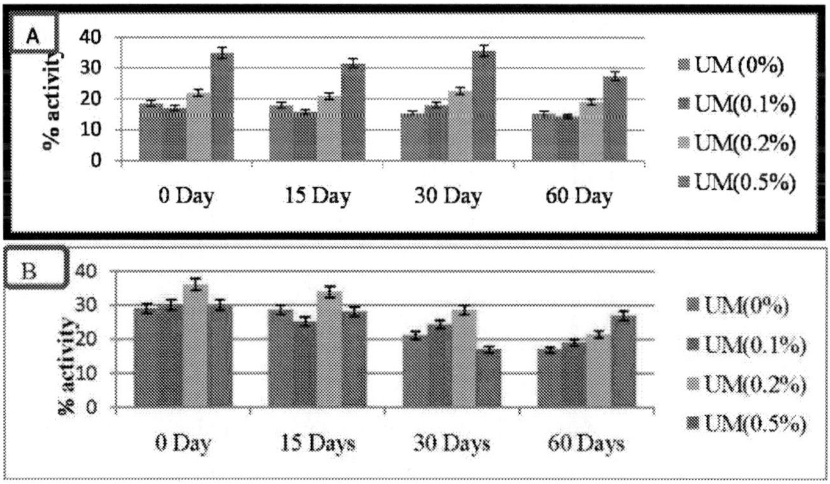

Figure 1. Free radical scavenging activities of edible oils and UM enriched edible oils (A: soybean oil and B: sunflower oil) by DPPH method.

Table 5. γ-oryzanol content of base oils and enriched oils with unsaponifiable matter (fatty alcohol-free)

Edible Oils + UMRBOFAD (Wt. %)	γ- oryzanol content (mg/100g)			
	0 Days	15 Days	30 Days	60 Days
SFO	-	-	-	-
SFO + UM (Wt.% 0.10)	3.24 ± 2.81	2.81 ± 0.10[a]	2.81 ± 0.18[a]	2.24 ± 0.28[a]
SFO + UM (Wt.% 0.20)	6.51 ± 0.35	6.72 ± 0.08[ns]	5.90 ± 0.10[a]	5.31 ± 0.09[a]
SFO + UM (Wt% 0.50)	16.19 ± 0.18	15.74 ± 0.14[a]	15.83 ± 0.07[b]	15.42 ± 0.29[a]
SBO	-	-	-	-
SBO + UM (Wt.% 0.10)	3.02 ± 0.06	2.85 ± 0.06[a]	2.38 ± 0.29[a]	2.84 ± 0.11[ns]
SBO + UM (Wt.% 0.20)	6.37 ± 0.46	5.39 ± 0.25[a]	5.27 ± 0.30[a]	4.82 ± 0.19[a]
SBO + UM (Wt.% 0.50)	16.58 ± 0.32	15.48 ± 0.42[a]	15.18 ± 0.45[a]	15.05 ± 0.60[a]

SFO = Sunflower oil, SBO = Soybean oil; UMRBOFAD = Unsaponifiable matter of rice bran oil fatty acid distillate. At $p < 0.05$, significant differences are shown by symbol (a-b). ns = Not significant.

Stability of Minor Components such as γ- Oryzanol during Storage Periods

The γ-oryzanol content of the individual oils and UM enriched oils was assessed periodically at room temperature for 60 days as shown in Table 5 the gamma oryzanol contents of UM enriched vegetable oils (SFO+UMR BOFAD;15.42 mg/100gm, SBO+UMRBOFAD;15.05 mg/100gm blended oil) were significantly higher than individual base oils (sunflower oil and soybean oil) during storage stage because sunflower oil and soybean oil do not content γ-oryzanol. The blended oils contain natural antioxidants, and therefore, these oils may supply good nutrition for human health.

Conclusion

The present study deals with the antioxidant activity of rice bran oil unsaponifiable matter isolated from rice bran oil fatty acid distillate when added to edible oils. The individual unsaponifiable matter components are γ – oryzanol, tocopherol, total phytosterol and squalene. The mixing of isolated unsaponifiable matters from RBOFAD in edible oils, namely, sunflower oil and soybean oil in different concentrations (Wt.%) such as 0.1%, 0.2%, 0.5%; show the slight difference of FFA, PV and *p*-anisidine values during 0 day to

60 days of storage of the oils. Antioxidant activities and γ-oryzanol content of UM enriched oils (SFO+UMRBOFAD; SBO+UMRBOFAD) are higher than individual oils (sunflower oil and soybean oil).

References

Aardt, Marleen van, Susan E. Duncan, Timothy E. Long, Sean F. O'Keefe, Joseph E. Marcy, and Susan R. Sims. 2004. "Effect of Antioxidants on Oxidative Stability of Edible Fats and Oils: Thermogravimetric Analysis." *Journal of Agricultural and Food Chemistry*, 52 (3): 587–91. https://doi.org/10.1021/jf030304f.

Al-Kharousi, Naima, Mohammed Al-Khusaibi, Ismail Al-Bulushi, Nejib Guizani, Mostafa Waly, and Mohammad Shafiur Rahman. 2016. "Physico-Chemical Characteristics and Oxidative Stability of Date-Pits Oil during Storage at Different Temperatures." *International Journal of Food Engineering*, 12 (4): 385–93. https://doi.org/10.1515/IJFE-2015-0173.

AOAC. 2005. *Official Methods of Ananlysis of AOAC International*, 18th edition (February): 20877–417.

Bhilwade, Hari Narayan, Naoto Tatewaki, Hiroshi Nishida, and Tetsuya Konishi. 2010. "Squalene as Novel Food Factor." *Current Pharmaceutical Biotechnology*., Vol. 11.

Chen, M.-H, and C J Bergman. 2005. "A Rapid Procedure for Analysing Rice Bran Tocopherol, Tocotrienol and g-Oryzanol Contents." *Journal of Food Composition and Analysis*, 18: 139–51. https://doi.org/10.1016/j.jfca.2003.09.004.

Chu, Yan-Hwa, and Yu-Lang Kung. 1998. "A Study on Vegetable Oil Blends." *Food Chemistry*, 62 (2): 191–95. https://ac.els-cdn.com/S0308814697002008/1-s2.0-S0308814697002008-main.pdf?_tid=50c8f986-7e64-4233-b956-a0831b25926d&acdnat=1527277638_db32d4c5660d57278caa5256a9eb0eec.

Ghosh, Mahua. 2007. "Review on Recent Trends in Rice Bran Oil Processing." *Journal of the American Oil Chemists' Society*, 84 (4): 315–24. https://doi.org/10.1007/s11746-007-1047-3.

Gomes, Tina A, Xuan Gu, Cindi R Brownmiller, Luke Howard, Cesar M Compadre, Sun-Ok Lee, and T A Gomes. 2015. "Effects of Tocols Rich Fraction Isolated from Rice Bran Oil Deodorizer Distillate on Plasma and Hepatic Lipid Concentrations in Rats." *Journal of Biosciences and Medicines*, 3 (3): 1–1. https://doi.org/10.4236/jbm.2015.39001.

Gupta, A. K., C. G. Savopoulos, J. Ahuja, and A. I. Hatzitolios. 2011. "Role of Phytosterols in Lipid-Lowering: Current Perspectives." *QJM*. https://doi.org/10.1093/qjmed/hcr007.

Hoed, V. Van, G. Depaemelaere, J. Vila Ayala, P. Santiwattana, R. Verhe, and W. De Greyt. 2006. "Influence of Chemical Refining on the Major and Minor Components of Rice Brain Oil." *Journal of the American Oil Chemists' Society*, 83 (4): 315–21. https://doi.org/10.1007/s11746-006-1206-y.

Lee, Ah Young, Myoung Hee Lee, Sanghyun Lee, and Eun Ju Cho. 2015. "Comparative Study on Antioxidant Activity of Vegetable Oils under *in Vitro* and Cellular System." *Journal of Agricultural Science*, 7 (3): p58. https://doi.org/10.5539/jas.v7n3p58.

Li, Hongyan, Ya-wei Fan, Jing Li, Liang Tang, Jiang-ning Hu, and Ze-yuan Deng. 2013. "Evaluating and Predicting the Oxidative Stability of Vegetable Oils with Different Fatty Acid Compositions." *Journal of Food Science*, 78 (4): H633–41. https://doi.org/10.1111/1750-3841.12089.

Nandi, Sumit, and Rupa Bhattacharyya. 2013. "Enzymatic Synthesis of Micronutrients Rich Diacylglycerol from Rice Bran Oil Fatty Acid Distillate." *International Journal of Science and Research (IJSR) ISSN*. Vol. 4. www.ijsr.net.

Nijjar, Prabhjot S, Frances M Burke, Annette Bloesch, Daniel J Rader, and F M Burke. 2010. "Role of Dietary Supplements in Lowering Low-Density Lipoprotein Cholesterol: A Review." *Journal of Clinical Lipidology*, 4: 248–58. https://doi.org/10.1016/j.jacl.2010.07.001.

Pal, Yogita P., and Amit P. Pratap. 2017. "Rice Bran Oil: A Versatile Source for Edible and Industrial Applications." *Journal of Oleo Science*, 66 (6): 551–56. https://doi.org/10.5650/jos.ess17061.

Rodrigues, Christianne E.C., Cintia B. Gonçalves, Elaine C. Marcon, Eduardo A.C. Batista, and Antonio J.A. Meirelles. 2014. "Deacidification of Rice Bran Oil by Liquid–Liquid Extraction Using a Renewable Solvent." *Separation and Purification Technology*, 132 (August): 84–92. https://doi.org/10.1016/J.SEPPUR.2014.05.009.

Sahu, S., M. Ghosh, and D. K. Bhattacharyya. 2020. "Utilization of Unsaponifiable Matter from Rice Bran Oil Fatty Acid Distillate for Preparing an Antioxidant-Rich Oleogel and Evaluation of Its Properties." *Grasas y Aceites*, 71 (1): a336–a336. https://doi.org/10.3989/GYA.0938182.

Sahu, S., M. Ghosh, D. K. Bhattacharyya, and D. K. Bhattacharyya. 2018. "Isolation of the Unsaponifiable Matter (Squalene, Phytosterols, Tocopherols, γ-Oryzanol and Fatty Alcohols) from a Fatty Acid Distillate of Rice Bran Oil." *Grasas y Aceites*, 69 (3): 262. https://doi.org/10.3989/gya.1112172.

Shiela, P. M., Y. N. Sreerama, and A. G. Gopala Krishna. 2004. "Storage Stability Evaluation of Some Packed Vegetable Oil Blends." *Journal of the American Oil Chemists' Society*, 81 (12): 1125. https://doi.org/10.1007/s11746-004-1029-x.

Smith, Theresa J. 2000. "Squalene: Potential Chemopreventive Agent." *Expert Opinion on Investigational Drugs*, 9 (8): 1841–48. https://doi.org/10.1517/13543784.9.8.1841.

Sohail, Muhammad, Allah Rakha, Masood Sadiq Butt, Muhammad Jawad Iqbal, and Summer Rashid. 2017. "Rice Bran Nutraceutics: A Comprehensive Review." *Critical Reviews in Food Science and Nutrition*, 57 (17): 3771–80. https://doi.org/10.1080/10408398.2016.1164120.

Troni, Kelly L, Simone M Silva, Antonio J A Meirelles, and Roberta Ceriani. 2013. "Study of Fatty Acid and Fatty Alcohol Formation from Hydrolysis of Rice Bran Wax." *Chemical Engineering Transactions*, 32 (292): 1747–52. https://doi.org/10.3303/CET1332292.

Yamamoto, Yukihiro, and Setsuko Har. 2012. "Novel Fractionation Method for Squalene and Phytosterols Contained in the Deodorization Distillate of Rice Bran Oil." In

Scientific, Health and Social Aspects of the Food Industry, 1st ed., 71–82. Japan: InTech. https://doi.org/10.5772/33126.

Yang, Dan-Xia, Xiao-Wei Chen, and Xiao-Quan Yang. 2018. "Phytosterol-Based Oleogels Self-Assembled with Monoglyceride for Controlled Volatile Release." *Journal of the Science of Food and Agriculture*, 98 (2): 582–89. https://doi.org/10.1002/jsfa.8500.

Chapter 14

Marker Assisted Breeding for Genetic Improvement of Bruchid Resistance in Pigeonpea, *Cajanus cajan* (L.): Achievements, Progress and Prospects

Sujit Kumar Mishra[1] and Jogeswar Panigrahi[2,*]

[1]Department of Zoology, School of Applied Sciences,
Centurion University of Technology and Management, Odisha, India
[2]Department of Biosciences and Bioinformatics,
Khallikote University, Berhampur, Odisha, India

Abstract

Pigeonpea (*Cajanus cajan* L.) is a protein enriched pulse crop and is domesticated mostly in the tropical and subtropical regions of the world. The net productivity of this crop is being affected by bruchid pests, especially the Callosobruchus spp. Host plant resistance against these pests paved a new way for eco-friendly and cost-effective breeding for bruchid resistance in this crop. For developing bruchid resistant cultivar, efforts have been conferred with screening of the pigeonpea genotypes for bruchid resistance. Some resistant genotypes were identified from the secondary and tertiary gene pool. After successful screening of 34 genotypes of pigeonpea, we at School of Life Sciences, Sambalpur University, Odisha, India, have selected *Cajanus cajanifolius* (ICPW-31) as the bruchid resistant genotype for hybridization with the identified susceptible cultivar, *Cajanus cajan* (ICPL 87119) to generate F_1-hybrids. Selfing of the F_1-hybrid resulted in the development of interspecific F_2 mapping population comprising of 170 F_2 individuals. Genotyping of the

* Corresponding Author's Email: drjpanigrahi@gmail.com.

In: Proceedings of BIOSPECTRUM
Editors: Sanket J. Joshi, Susmita Mukherjee and Moupriya Nag
ISBN: 978-1-68507-985-7

mapping population with SSR, SCoT and CDDP markers, phenotyping of the F_3 seed families for bruchid resistance and development of genetic linkage map are under progress. Once the genetic linkage map has been constructed and marker-trait associations have been established, the tightly linked flanking marker(s) for the bruchid resistance trait could be validated and precisely utilized in marker assisted breeding programs (MAS) for development of bruchid resistance pigeonpea variety.

Keywords: pigeonpea, marker assisted breeding, bruchid pest management, hybrid seed development

Introduction

Pigeonpea is a warm-season perennial legume crop cultivated mostly in the subtropical and tropical areas of the world including Asia, Latin America, Caribbean and Africa. It is economically much important and holds an important position in human nutrition due to richness in proteins aided with appropriate amount of carbohydrates, minerals and vitamins. Pigeonpea also play a crucial role in enhancing fertility of the soil by forming a symbiotic relationship with root nodule bacteria. This adds organic matter and enhances phosphorus and nitrogen in the soil [13]. The annual production of pigeonpea over the globe (as in 2016) is about 4.49 million tonnes from land area of about 5.41 million hectares with yield of 829 kg per hectare [7]. India is the largest producer of pigeonpea and has 5.06 million hectares land under pigeonpea cultivation. India alone contributes about 67.7% of the world production, with average yield of 6499 Hg/Ha [6]. Particularly in India, pigeonpea accounts for about 20 percent of total pulse production in the world [6]. The net productivity of pigeonpea is being constrained with various biotic and abiotic stresses associated with less efficient management and cultivation practices [23]. Among the biotic stresses, the most detrimental storage insect pests-bruchids belongs to the genus Callosobruchus (*C. chinensis* L. and *C. maculatus* F.) are becoming serious threat to pigeonpea production and causing substantial post-harvest damage to pigeonpea seeds during storage conditions [8, 13, 17]. Utmost level of damage observed mainly in the subtropical and tropical countries of the world due to high level of relative humidity and temperature [10]. The bruchid pests invade matured pods and seeds of pigeonpea in the storage condition and to some extent in the farming field, and eventually diminish the nutritional quality and economic importance of the seeds. The biotic stresses on field followed by loss on storage conditions due to bruchids caused a huge shortage of pigeonpea grains in India during the last couple of years. As a consequence, the price has been escalated and the split grains has become unaffordable to a large segment of consumers for whom these grains are the prime source of protein supplements. To discourage the growth of these pests in pigeonpea, several management tactics including chemical, cultural, biological, botanical and physical control methods have been employed since decades by manipulating the host environment [13]. However, none of these methods are found successful for management of these pests especially in pigeonpea. Though chemical pesticides are mostly preferred by farmers, these are found to be detrimental as are not eco-friendly and concern with food safety and several health issues. An alternative and eco-friendly control tactic against these pests in pigeonpea could be realized by restoring the genetic resistance of the host through amalgamation of conventional breeding process with modern breeding tools. This could be possible through exploitation/utilization of host plant resistance sources identified in the secondary and tertiary gene pool of pigeonpea. To this end, this paper enumerated the outstanding achievements, current progress and future prospects of marker assisted breeding (MAB) for genetic improvement of bruchid resistance in pigeonpea.

Needs of Breeding Pigeonpea for Bruchid Resistance

The use of host plant resistance sources against bruchids in breeding pigeonpea has tremendous advantage over the other control measures, especially the most preferred chemical pesticides, available for management of these pests. Cultural control methods are effective but are species-specific and require long-term preparation and appropriate timing for treatment [9, 13]. Physical control techniques such as application of inert dusts and radiations are costly and lead to seed viability loss [13]. Although, use of biological control agents by small and marginal farmers is somehow effective but, it has some drawbacks associated with periodic supply of nutrients to the control agents and appropriate timing of release of these agents to the storage environment at appropriate stages of larval development [9]. Application of chemical pesticides has a thoughtful effect on management of these pests but it was found that different biotypes of bruchids showed resistance against the chemical pesticides [5, 22] and their use at elevated dose lead to the toxic accumulation in the food product. Moreover, chemical pesticides are associated with expensiveness, severe health hazards, risk of food safety and pollution in the environment [10]. The disadvantage associated with these methods has created a worldwide interest in utilization of host plant resistance sources available in different gene pool of pigeonpea in breeding for bruchid resistance. Once resistant cultivar against these pests becomes developed, the small and marginal resource-poor farmers would be benefited by adopting it, as the use of this cultivar could bear little additional cost. Breeding cultivar resistant to bruchid pest has another benefit in sense that the existing technological knowledge and practices have the farmers may require little alteration while adopting the newly developed resistant cultivar [4]. Further, exploitation of host resistance in breeding pigeonpea for bruchid resistance could be compatible with additional pest management practices including biological, cultural, physical and chemical control tactics, and hence it is established as a vital component of integrated pest management (IPM) program [15]. The need of breeding pigeonpea by exploiting host resistance is, therefore, imperative as far as the sustainable productivity and food safety is concern.

Challenges in Breeding Pigeonpea for Bruchid Resistance

Breeding pigeonpea for resistance to bruchid is a complicated process with various challenges. Global warming associated with accelerating climate volatility emerged as a major problem for the pigeonpea breeders to breed for bruchid resistance cultivar in pigeonpea. Ever changing climatic conditions lead to changes in the biotype, agro-climatic zones, geographical distribution and population dynamics of the insect pests around the world [1, 16]. Development of a resistance cultivar against multiple biotypes of bruchid, considering entire agro-climatic zones would be a great challenge to the pigeonpea breeders. Environment also plays a key role for the growth, flowering as well as pod setting in pigeonpea [12]. Hence, breeding for a wider environmental adaptation with contrasting host resistance against bruchids is a complex issue to be resolved. Besides this, attaining resistance to bruchids in pigeonpea without neglecting any agro-economic trait is a cumbersome task being faced by the pigeonpea breeders since unwanted genetic linkage(s)/linkage drag(s) impelled co-inheritance and co-expression of undesirable and desirable traits in a single cultivar [10]. It is, therefore, a great challenge to the pigeonpea breeders to obtain a perfect and stable cultivar, which would be bruchid resistant and not be deviating from consistent productivity, high yield, good seed quality, better seed size and higher nutritional value. Moreover, there is possibility that the unnecessary genetic linkages between the desirable and undesirable trait(s) could be broken out at the time of meiosis i.e., during crossing over between the homologous chromosomes [26]. However, this requires growing of a greater number of F_2 plants for rapid release of the new recombinants with desired traits. Further, backcrossing could also decrease the unusual effects of unwanted linkages or linkage drags [10]. Although pigeonpea has a rich reservoir of genes related to various stresses and yield components in its wild gene pools, lack of interspecific cross-compatibility is also a big issue to be tackled [12].

Genetic Resources and Mapping Population Presently Available to Breed for Bruchid Resistance in Pigeonpea

Successful breeding program relies on the extent of genetic polymorphism among the crossable parental lines, heritability of the desirable trait and the intensity of selection pressure exercised (Falconer, 1989). Therefore, the expected genetic gain for a desired trait is dependent on accurate selection of parents for the particular trait. To this end various genotypes belonging to the primary, secondary and tertiary gene pool of pigeonpea have been evaluated to obtain genetic resources of host resistance against the bruchid pests [8, 14]. Some accession of *C. Cajanifolius* (ICPW-30, 31) *C. scarabaeoides* (ICPW-94, 95) and *C. platycarpus* (ICPW-67, 68) were identified as the resistance sources to bruchid pests, *Callosobruchus chinensis* and *C. maculatus* [13]. These genotypes are to be exploited in breeding programs for transfer of the resistance traits to the cultivated backgrounds. It was also observed that the cultivars placed in the primary gene pool of pigeonpea are more vulnerable to these pests than the wild land racers in the secondary and tertiary gene pool. This finding has been supported by the earlier results obtained through screening of various legume crops like *Vigna radiata* and *V. mungo* [3, 11, 18, 19, 20, 21].

Mapping populations are the source material for construction of genetic linkage maps. Identification of SITLs (simple inherited trait loci) and QTLs (quantitative trait loci) for the desired traits is based on the mapping populations. In this connection, more than 30 mapping populations for various abiotic and biotic stresses in pigeonpea have been developed worldwide [2, 25]. But there is dearth of literature available on development of mapping populations concerning resistance to bruchid pests in pigeonpea. We at School of Life Sciences, Sambalpur University developed interspecific mapping population of 170 F_2 individuals involving *C. cajan* (ICPL-87119) and *C. cajanifolius* (ICPW-31) as susceptible and resistant parent respectively. This population could successfully be utilized for constructing genetic linkage maps and identifying SITLs/QTLs linked to the bruchid resistance trait(s).

Progress and Future Prospects in MAB for Genetic Improvement of Bruchid Resistance in Pigeonpea

Marker assisted breeding (MAB) refers to the utilization of DNA markers in breeding for developing agro-economically desired crop i.e., superior lines for the desired traits. It relies on the identification of genetic resource(s) with desired trait(s), development of mapping population and molecular markers, establishment of marker(s)-trait/gene linkage, validation of linked markers in diverse genetic backgrounds and utilization of the linked markers in conventional breeding for marker assisted selection and breeding (MASB). Various studies were carried out in different pulses during the last several years to identify the DNA markers linked to either SITLs or QTLs conferring bruchid resistance [13]. However, MAB for genetic improvement of pigeonpea for bruchid resistance is at its incipient phase because of unavailability of DNA markers associated with bruchid resistance trait(s). Development of genetic linkage map using the genotypic data obtained from SSR, SCoT and CDDP marker analyses in F_2 mapping population and phenotypic trait data obtained from the F_3 seed family is under progress.

Once the marker-trait associations have been established through genetic linkage mapping, the linked marker(s), if found tightly linked to the bruchid resistance trait(s), could be validated in different genetic backgrounds for further utilization in MAB program. The introgression of the resistance gene(s)/trait(s) through MAB programs could increase the yield potential of this crop by eliminating the loss incurred due to bruchid pests. Further, if the flanked DNA marker would be distantly apart from the resistance trait(s), the map has to be enriched with more DNA markers. Besides this, a pretty good number of simple sequence repeats (SSR) and single nucleotide polymorphism (SNP) markers were developed during the last couple of years [24]. The utilization of these markers for mapping bruchid resistance loci and subsequent MAB in pigeonpea is under progress.

Conclusion

Pigeonpea has a tremendous impact on food security of the growing population of the world. However, the net output of this crop is harshly affected by the bruchid pests. Several control tactics including cultural, biological and chemical control methods have been put forth into practices for controlling these bruchid pests. These methods were found significantly less successful for managing these pests. In this context, the potential of MAB involving host plant resistance has been appreciated in developing bruchid resistance pigeonpea cultivars. Development and successful deployment of cultivars resistant to bruchids would result in lesser utilization of chemical pesticides and provide sustainable yield by culminating losses. From this viewpoint, momentous progress has been attained in identifying resistance genetic resources, developing mapping population involving bruchid resistance and molecular markers. Efforts are now focused on construction of genetic linkage map using genotypic and phenotypic trait obtained from the mapping population and identification of flanking marker associated with the bruchid resistance. The identified linked DNA markers could be validated and utilized in MASB programmes for development of bruchid resistance cultivar(s).

In brief, for developing cultivar(s) resistant to bruchid pests, there is need:

- To understand the underlying mechanisms behind host plant resistance of wild accessions to bruchid pests.
- To identify the biochemical traits imparting resistance against bruchid pests in the existing genetic resources.
- To diversify the genetic base of primary gene pool and increase the levels of pest resistance by utilizing cross compatible wild relatives in breeding program.

- To identify SITLs/QTLs linked to bruchid resistance and its constituent traits, and pyramiding the SITLs/QTLs into a single genetic setting to augment the durability of resistance.
- To validate the identified linked markers in different genetic background prior to their utilization in MAB.
- To deploy MAB and gene pyramiding strategies to unite the alleles controlling resistance to bruchid and other biotic and abiotic stresses.

References

[1] Bahl, P. N., 2015, Climate change and pulses: approaches to combat its impact. *Agric Res,* 4, 103-108.

[2] Bohra, A., Saxena, R. K., Saxena, K. B., Sameerkumar, C. V., Varshney, R. K., 2014, Advances in pigeonpea genomics, In: *Legumes in the Omic Era*, edited by S. Gupta, N. Nadarajan, D. Sen Gupta (New York; Heidelberg; Dordrecht; London: Springer), pp. 95-110.

[3] Chen, H. M., Chien, A. L., Kuo, C. G., Ching, M. C., 2007, Development of a molecular marker for a bruchid (Callosobruchus chinensisL.) resistance gene in mungbean. *Euphytica,* 157, 113-122.

[4] Edmeades, G. O., Bolanos, J., Banziger, M., Ribaut, J. M., White, J. W., Reynolds, M. P., Lafitte, H. R., 1998, Improving crop yields under water deficits in the tropics. In: Crop productivity and sustainability-shaping the future, edited by V. L. Chopra, R. B. Singh, A. Varma. *Proceedings of the 2nd International Crop Science Congress,* (New Delhi: Oxford and IBH), pp. 437-451.

[5] Egwuatua, R. I., 1987, Current status of conventional insecticides in the management of stored product insect pests in the tropics. *International Journal of Tropical Insect Science,* 8, 695-701.

[6] FAOSTAT, 2016; Date of access: https://10/01/2018; faostat3.fao.org/compare/E.

[7] FAOSTAT, 2018; Date of access: https://10/01/2018; http://faostat.fao.org.

[8] Jadhav, D. R., Mallikarjuna N., Rao, G. V. R., 2012, Callosbruchus maculatus resistance in some wild relatives and interspecific derivatives of Pigeonpea. *Indian Journal of Plant Protection*, 40, 40-44.

[9] Kananji, G. A. D., 2007, *A Study of Bruchid Resistance and its Inheritance in Malawian Dry Bean Germplasm* [Ph.D. Thesis.] African Centre for Crop Improvement (ACCI), School of Biochemistry, Genetics, Microbiology and Plant Pathology, Faculty of Science, Agriculture University of KwaZulu Natal Republic of South Africa, KwaZulu-Natal, South Africa).

[10] Keneni, G., Bekele, E., Getu, E., Imtiaz, M., Damte, T., Mulatu, B., Dagne, K., 2011, Breeding food legumes for resistance to storage insect pests: potential and limitations. *Sustainability,* 3, 1399-1415.

[11] Ku-Hwan, L., Seung-Keun, J., Sang, P. II., Beom-Heon, S., Yong-Gu, C., Hong-Sig, K., 2002, Agronomic characteristics of resistant mungbean to *Callosbruchus chinensis*. *Journal of Agricultural Science,* 19, 41-52.

[12] Magadum, S. K., Duggi, S., Gangapur, D., Verma, S. K., 2013, Role of molecular breeding in geneti improvement of pigeonpea. *International Journal of Bio-resource and Stress Management* 4, 382-388.

[13] Mishra, S. K., Macedo, M. L. R., Panda, S. K., Panigrahi, J., 2018, Bruchid pest management in pulses: past practices, present status and use of modern breeding tools for development of resistant varieties. *Ann Appl Biol,* 172(1), 4-19.

[14] Mishra, S. K., Panigrahi, J., 2017, Identification of inter-specific hybrid between *Cajanus cajan* (L.) and *C. cajanifolius* (H.) using cyto-morphological and DNA markers. *Electronic Journal of Plant Breeding*, 8(1), 63-71.

[15] Panda, N., Khush, G. S., 1995, *Host Plant Resistance to Insects*. CAB International in association with International Rice Research Institute (IRRI), (Guildford, UK: Biddles Ltd.).

[16] Sharma, H. C., 2016, Climate change vis-a-vis pest management, In: *Proceeding of Conference on National Priorities in Plant Health Management,* (Plant Protection Association of India, NBPGR, Hyderabad, India).

[17] Singh, U., Jambunathan, R., 1990, Pigeon pea post-harvest technology. In: *The Pigeon Pea*, edited by Y. L. Nene, S. D. Hill, V. K. Sheila. CAB Press, Wallingford, UK, pp. 435-455.

[18] Somta, C., Somta, P., Tomooka, N., Ooi, P. A. C., Vaughan, D. A., Srinives, P., 2008, Characterization of new sources of mungbean (Vigna radiata (L.) Wilczek) resistance to bruchids, Callosobruchus spp. (Coleoptera: Bruchidae). *Journal of Stored Products Research,* 44, 316-321.

[19] Somta, P., Ammaranan, C., Ooi, P. A. C., Srinives, P., 2007, Inheritance of seed resistance of bruchids in cultivated mungbean (Vigna radiata L. Wilczek). *Euphytica,* 155, 47-55.

[20] Somta, P., Talekar, N. S., Srinives, P., 2006, Characterization of *Callosobruchus chinensis* (L.) resistance in *Vigna umbellate* (Thunb.) Ohwi & Ohashi. *Journal of Stored Products Research,* 42, 313-327.

[21] Souframanien, J., Gupta, S. K., Gopalakrishna, T., 2010, Identification of quantitative trait loci for bruchid (Callosobruchus maculatus) resistance in black gram [Vigna mungo (L.) Hepper]. *Euphytica,* 176, 349-356.

[22] Talukder, F., 2009, Pesticide resistance in stored product insects and alternative biorational management: a brief review. *Agricultural and Marine Sciences,* 14, 9-15.

[23] Varshney, R. K., Chen, W., Li, Y., Bharti, A. K., Saxena, R. K., et al., 2012, Draft genome sequence of pigeonpea (*Cajanus cajan*), an orphan legume crop of resource-poor farmers. *Nature Biotechnology,* 30, 83-89.

[24] Varshney, R. K., 2016, Exciting journey of 10 years from genomes to fields and markets: Some success stories of genomics-assisted breeding in chickpea, pigeonpea and groundnut. *Plant Science*, 242, 98-107.

[25] Varshney, R. K., Penmetsa, R. V., Dutta, S., Kulwal, P. L., Saxena, R. K., Datta, S., et al., 2010, Pigeonpea genomics initiative (PGI): an international effort to improve crop productivity of pigeonpea (*Cajanus cajan* L.). *Mol Breed.*, 26, 393-408.

[26] War, A. R., Murugesan, S., Boddepalli, V. N., Srinivasan, R., Nair, R. M., 2017, Mechanism of Resistance in Mungbean [*Vigna radiata* (L.) R. Wilczek var. *radiata*] to Bruchids, *Callosobruchus* spp. (Coleoptera: Bruchidae). *Front. Plant Sci.,* 8, 1031, doi: https://10.3389/fpls.2017.01031.

Chapter 15

A Comparative Analysis of the Role of Different Agents in Preventing Browning of Banana (*MusaSp.*)

Susmita Mukherjee[1], Tirtha Tarafdar[1], Bhaskarjyaa Chatterjee[1], Binapani Giri[2] and Sonali Paul[1,*]

[1]Department of Biotechnology,
University of Engineering and Management,
University Area, Kolkata, West Bengal, India
[2]Department of Biotechnology,
Institute of Genetic Engineering,
Kolkata, West Bengal, India

Abstract

Banana (*Musa sp.*) is widely cultivated for its high minerals and vitamin content which help in maintenance of a healthy life. But one very important limitation in consumption of these ripe bananas is difficulty in their transportation and storage for long period of time. This is because they tend to brown very easily due to the activity of polyphenol oxidase, an enzyme found in banana, which leads to reduction in taste, appearance and subsequent market losses. In our work we studied the effect of treatment of banana with various natural and chemical agents. We found that apart from β-cyclodextrin which is a known chemical inhibitor of polyphenol oxidase, refrigeration as well as treatment with lemon juice, green tea was able to prevent browning to the greatest extent. These

---

* Corresponding Author's Email: sonali.paul@uem.edu.in.

In: Proceedings of BIOSPECTRUM
Editors: Sanket J. Joshi, Susmita Mukherjee and Moupriya Nag
ISBN: 978-1-68507-985-7

results hold good promise in lemon juice, green tea being utilized as common, cheap anti browning agents instead of chemicals in places where refrigeration is not available. This will ensure the maintenance of ripe banana in its original state for a longer time.

Keywords: Banana, refrigeration, lemon juice, green tea, β-cyclodextrin, polyphenol oxidase, browning

Introduction

Banana is one of the highly commonly consumed fruit in many parts of the world due to its very high nutritional value, easy cultivation methods and inexpensiveness, which makes it a fruit of choice for people of all economic and social classes. Availability throughout the year and easy consumption method makes it one of the most acceptable fruits across all geographic regions and all ages. However, use of banana is strongly limited by its fast ripening rate which is expressed as external and internal browning and, consequently, decreases the acceptability of the fruit by consumers. This phenomenon typically occurs in cells from ripe fruits, where the loss of compartmentalization allows the contact between phenols and their oxidative enzymes. [1]

Banana is an important source of polyphenols. Polyphenols are natural antioxidants. Dietary phenolics or polyphenols constitute one of the most widely distributed groups of bioactive molecules in the plant kingdom. More than 8000 phenolic molecules are known, and among them, over 4000 flavonoids have been characterized [2, 3, 4].

Polyphenol oxidase (PPO) is an enzyme which is found in banana [5]. It is a nuclear encoded enzyme that catalyzes the oxidation of polyphenols forming corresponding quinone intermediates that polymerize to form the coloured pigment [6]. Two types of the oxidative reaction are catalyzed: the hydroxylation of monophenols to O- diphenols and the oxidation of O-diphenols to O-quinones, leading to black or brown pigment formation. Browning of banana causes negative effects in taste, appearance and subsequent deterioration of economic significance of banana.

The quality of fruits, depends on controlling the enzymatic browning that affects its sensory and economical qualities. Consumer demand for minimally processed and fresh-like food products has undergone a continuous rise in recent years. This has led to the necessity for evaluation of the inhibitory effect of natural compounds on PPO activity and its applicability to the food industry.

Many methods were previously applied to control browning of banana [7]. In our present work we have studied the role of different agents in preventing browning of banana Addition of chemicals, natural ingredients, refrigeration, are among the most effective methods for the inhibition of polyphenol oxidases (PPO).

Methods

Selection of Sample and Preparation of Sample Extracts

Ripe banana fruits of Musa acuminata-AB cultivar group and AAB cultivar group which are common in the Indian subcontinent were taken.0.5g of both pulp and peel were taken as study samples. The samples were taken, boiled, filtered and filtrate was taken.

Methods Employed for Treated and Untreated Samples

The samples were kept at room temperature and also subjected to refrigeration (4°C), treatments with different natural items – orange juice (50%), lemon juice (50%), soda water (25%), green tea (3 g/l), chemicals - β-cyclodextrin (1%), propyl gallate (0.1%). All treatments were done for 7 days

The banana pulp and peel samples were kept at room temperature for 7 days.

Determination of Total Polyphenol

5.0 mL of extract were placed in calibrated flask and filled up to 25 mL by distilled water. 1.0 mL of Folin-Ciocalteu reagent, 10.0 mL of distilled water were added to 2.0 mL of diluted extract (from 25 mL calibrated flask) and filled up to 25 mL by sodium carbonate solution (290 g/L). An absorbance of prepared sample was measured by means of UV spectrophotometer at 760 nm after 30 minutes of incubation in darkness. All determinations were performed in triplicate. The content of polyphenols was calculated from the formula:

$$X = (62.5 \times A1 \times m2) / (A2 \times m1)$$

where A1 – absorbance of pyrogallol solution, A2 – absorbance of investigated samples, m1 – mass of investigated sample [g], m2 – mass of pyrogallol [g].[8]

Solution of Reference Substance (Pyrogallol)

50 mg of pyrogallol was dissolved in distilled water and filled up to 100 mL in calibrated flask. 5.0 mL of obtained solution was diluted in another 100 mL calibrated flask. Absorbance of 2.0 mL of pyrogallol solution (with the adequate reagents) was measured by the same method as described for samples.

Results

Table1. Total polyphenol content in sample with varied treatment

Sample	Total Polyphenol Content							
	Untreated	Refrigeration	Lemon Juice	Orange Juice	Soda Water	Beta Cyclodextrin	Propyl Gallate	Green Tea
Musa acuminate AB group peel	68	23	25	60	60	30	50	34
Musa acuminata AB group pulp	60	20	28	38	40	28	45	30
Musa acuminata AAB group peel	65	23	28	40	40	36	32	26
Musa acuminata AAB group pulp	60	23	30	35	32	30	34	32

Visual Observation

The banana pulp and peel pieces which were kept at room temperature showed the greatest amount of darkening. The samples which were kept at 4°C and treated with lemon juice, green tea, β-cyclodextrin showed the least browning.

Total Polyphenol Content

There were differences in the total polyphenol content in different varieties of banana as well as in different parts (Table 1).

Discussion

Difference in Polyphenol Content in Different Varieties and Fruit Parts

Polyphenol content has been found to vary among different varieties, different parts of banana [9]. In our study too, we have found differences in polyphenol content among Musa acuminata AB, AAB group and among fruit peel, fruit pulp. This is in confirmation with other previous observations [9].

Difference in Polyphenol Oxidase Activity

The samples were treated under various conditions and showed variations in colouration. The samples which were kept at 4°C showed the least browning effect. This is because polyphenol oxidase (PPO) activity is inhibited at 4°C [10]. PPO shows high activity at 30°C [11]. This was the existing room temperature at which maximum browning was observed in the untreated sample. PPO shows maximum activity at pH 6.5 – 7.0 with least activity detected at pH 2.5 and 9.5 [11]. Polyphenols treated at pH between 4-7 (orange juice, soda water) showed higher degree of retention. This is in confirmation with similar other observations [12]. This correlated with higher degree of browning observed in samples treated with orange juice and soda water. Lemon juice has a pH of 2.0. Polyphenols are less stable at lower pH [13]. This is in confirmation with treatment of the samples with lemon juice (pH 2.0) which showed least amount of browning. Green tea extract has been found to effectively retard the browning of banana. Green tea extract which is rich in flavonoids has an inhibitory effect on PPO activity [14]. Our work showed the low browning effect when samples were treated with green tea extract. Cyclodextrins trap the PPO polyphenolic substrates, making them unavailable to PPO oxidation thereby inhibiting PPO activity [15]. This is in confirmation with our observations of low browning effect observed in samples treated with β-cyclodextrin. Propyl gallate is an antioxidant. It protects against oxidation by hydrogen peroxide and oxygen free radicals [16]. This agrees with our work where we have observed lessening of PPO activity in samples treated with propyl gallate especially in Musa acuminata AAB cultivar.

Conclusion

Different agents – both natural and chemical can be utilized to increase the shelf life of banana. Due to its high efficiency, low cost, nontoxicity and least health hazards natural agents like lemon juice, green tea extract apart from refrigeration holds good promise for being utilized to prevent browning of banana.

References

[1] Su X, Jiang Y, Duan X, Liu H, Li Y, Lin W, Zheng Y. (2005). Effects of pure oxygen on the rate of skin browning and energy status in longan fruit. *Food Technol Biotechnol*. 43(4). 359–365.

[2] Harborne J. B., Williams C. A. (2000). Advances in flavonoid research since 1992. *Phytochemistry*. 55 (6). 481-504.

[3] Bravo L. (1998). Polyphenols: chemistry, dietary sources, metabolism and nutritional significance. *Nutr. Rev*. 56 (11). 317-333.

[4] Cheynier V. (2005). Polyphenols in foods are more complex than often thought. *Am. J. Clin. Nutr.* 81 (1 Suppl.). 223S–229S

[5] Escalante-Minakata P, Ibarra-Junquera V, Ornelas-Paz JJ, García-Ibáñez V, Virgen-Ortíz JJ, González-Potes A, Pérez-Martínez JD, Orozco-Santos M. (2018). Comparative study of the banana pulp browning process of 'Giant Dwarf' and FHIA-23 during fruit ripening based on image analysis and the polyphenol oxidase and peroxidase biochemical properties. *3 Biotech*. 8(1). 30

[6] Aziz G.M. and Al-Sa'ady A.J.R. (2017). Extraction conditions of polyphenol oxidase from banana peel. *Baghdad Science Journal*. 13(3). 469-474.

[7] Almeida M.E. and Noquiera J.N. (1995). The control of polyphenol oxidase activity in fruits and vegetables. A study of the interactions between the chemical compounds used and heat treatment. *Plant Foods Hum Nutr*. 47(3):245-256

[8] Modnicki D, Balcerek M. (2009). Estimation of total polyphenols contents in Ocimum basilicum L., Origanum vulgare L. and Thymus vulgaris L. commercial samples. *Herba Polonica*. 55(1), 35-42.

[9] Fatemeh, S. R., Saifullah, R., Abbas, F. M. A. and Azhar, M. E. (2012). Total phenolics, flavonoids and antioxidant activity of banana pulp and peel flours: influence of variety and stage of ripeness. *International Food Research Journal*. 19 (3): 1041-1046

[10] Xu J. (2005). The effect of low-temperature storage on the activity of polyphenol oxidase in *Castanea henryi* chestnuts. *Postharvest Biology and Technology*. 38 (1). 91-98

[11] Mizobutsi G.P., Finger F.L.; Ribeiro R. A.; Puschmann R; Neves L.L.M.; Mota W. F. (2010). Effect of pH and temperature on peroxidase and polyphenoloxidase activities of litchi pericarp. *Sci. Agric*. 67 (2). 213-217.

[12] Sun H.N., Mu T.H.and Xi L.S. (2017). Effect of pH, heat, and light treatments on the antioxidant activity of sweet potato leaf polyphenols. *International Journal of Food Properties*. 20 (2). 318-332.

[13] Ruenroengklin N, Zhong J, Duan X, Yang B, Li J, Jiang Y. Effects of various temperatures and pH values on the extraction yield of phenolics from litchi fruit pericarp tissue and the antioxidant activity of the extracted anthocyanins. (2008). *Int. J. Mol. Sci*. 9(7). 1333-1341.

[14] Klimczak I, Gliszczynska-Swiglo A (2017). Green tea extract as an anti-browning agent for cloudy apple juice. *J. Sci. Food Agric*. 97(5). 1420-1426.

[15] de la Rosa L.A Mercado-Mercadoa G., García J.R., González-Aguilarb G.A., Alvarez-Parrillaa E. (2010). Peach polyphenol oxidase inhibition by b-cyclodextrin and 4-hexylresorcinol is substrate dependent. *Cyta J. Food*. 8(2). 87-93.

[16] Lin YF, Hu YH, Lin HT, Liu X, Chen YH, Zhang S, Chen QX (2013). Inhibitory effects of propyl gallate on tyrosinase and its application in controlling pericarp browning of harvested longan fruits. *J Agric. Food Chem*. 61 (11). 2889-2895.

Chapter 16

Micropropagation of Banana: A Promise of High Quality Production and Improved Food Security

Udayan Bhattacharyya, Ahana Sengupta, Manami Adhikari, Saranya Ganguly, Tridha Deb and Basumita Roychowdhury*

Department of Biotechnology,
Bengal Institute of Technology,
Kolkata, West Bengal, India

Abstract

Bananas (*Musa sp.*) are considered the most important fruit crop after mango in India. The major problem facing banana cultivation is the method of vegetative propagation and disease-free planting material. This difficulty can be solved by employing commercial tissue culture technology for the mass production of the popular banana varieties. A complete micropropagation protocol was developed with MS medium with growth regulators BAP and NAA (4 mg + 2.5 mg/l). After regeneration of multiple shoots, each seedling was reared in rooting medium followed by primary and secondary hardening. These hardened plants have successfully established themselves in the soil.

Keywords: micropropagation, high yield generation, culture medium, growth index

---

* Corresponding Author's Email: basumitarc@gmail.com.

In: Proceedings of BIOSPECTRUM
Editors: Sanket J. Joshi, Susmita Mukherjee and Moupriya Nag
ISBN: 978-1-68507-985-7

Introduction

Banana is a perennial herbaceous monocot that belongs to the genus Musa of the Musaceae family which is considered to be one of the oldest fruits [3]. Cultivation can be done in subtropical conditions; if the planting time is regulated in a particular way, i.e., the batches are initiated in summer, short in autumn and ripened in winter [1, 9, 10]. Its genesis occurred in Malaysia where the process followed a complex hybridization. Banana (*Musa sp.*) In India is the second most important fruit after mango. Banana is the most popular and preferred fruit among all classes of people for its various qualities, viz. year round availability, affordable price, varietal range, taste, nutritional and medicinal value. It is basically cultivated in the tropics, and grows best between 13°C and 38°C room temperature with relative humidity between 75% and 85% [8].

In the current scenario, tissue culture is preferred over the conventional selection method. Micropropagation of bananas gives much faster yields and healthy growth and a shorter period for the production cycle is also observed [2, 4]. This saves time and multiplies the batches of crops several times a year, whatever the season. Although this is a complex and expensive process and there is also a huge risk of contamination, the advantages outweigh the disadvantages of tissue culture [8]. As it successfully resulted in the production of a good no. of disease-free plants from a single plant tissue, its yield was estimated to be 39% higher than the conventional method. [5, 6, 8].

Auxin, cytokinin, gibberellins and abscisic acid such as kinetin, indole-3-acetic acid, benzylaminopurine, and these growth regulators are used for *in vitro* regeneration of sample plants. To reduce apical meristem dominance and to generate multi-shoot cytokinins, 6-benzylaminopurine (BAP), Kinetin (kin), N6-(2-isopentenyl)adenine (2iP) can be used to trigger root auxins, to know. Indole-3-butyric acid (IBA), α-Naphthaleneacetic acid (NAA) [2, 6, 8] are also part of the list. It has been noted that among the different mechanisms of banana cultivars, BAP has been shown to be the most effective compared to other cytokinins in inducing the generation of multiple shoot tips [2, 10]. The use of BAP in addition to basic media (Murashige and Skoog, 1962) has been shown to be beneficial in most established banana apex cultivation systems, as stimulated growth of axillary and adventitious buds has been observed [7, 8].

Banana cultivation scenario [2, 11]:

- Globally the production of banana is about 97.5 million.
- The livelihood of millions of people in India is supported by it.
- The total annual production is 16.91 million tons from 490.70 k ha, with a national average of 33.5 T/ha of area. It contributes 37% to total fruit production.
- Jalgaon located in Maharashtra is considered to be the major banana growing district. About 50,000 hectares of area is used for the cultivation and most of this crop is grown by planting sucker.

Thus, the above information gives a clear idea of why tissue culture needs to merge with conventional banana breeding. This would not only save time, but also meet the vast demand for saplings among farmers, thus supporting the livelihoods of many and helping the country's economy grow faster.

Material and Methods

Table 1. Composition of medium found after laboratory standardization gives best response

Reagents	Concentration
MS Media	4.4 gm
BAP	4 ml
NAA	0.25 ml
Sodium Dihydrogen Phosphate	0.488 gm
Adenine Sulphate	0.05 gm
Sucrose	30 gm
Agar	8 gm
Activated Charcoal	

MS medium is used for the cultivation of bananas in the laboratory. For the preparation of 1 liter of MS medium, the following reagents are added and diluted with sterile distilled water. The pH was maintained at 5.8. NAA and BAP are included for the morphogenesis and development of banana tissue. To regulate the acidity of the system, sodium dihydrogen phosphate is used in the medium [11]. Here, sucrose is used as a carbon source. Agar is used as a gelling agent. To avoid the darkening of the tissues, due to the oxidation of the polyphenolic compounds discharged from the injured tissues, activated carbon is used. Finally, the prepared medium is autoclaved at 121°C. for 30 minutes.

The explant is a piece of plant tissue used in tissue culture. Explants are isolated from the tissues of higher plants and then used in culture [10, 11]. Any other means than asexual propagation are not feasible for banana production because the seeds are not viable. The banana tree has a real stem called a rhizome, which bears several buds. Its own pseudo-stem forms as each of these buds proliferate. These daughter plants are called suckers. The lateral shoot that develops from the rhizome is known to be a sucker. This emerges from the mother plant. It is one of the types of asexual synthesis that makes the cultivation of banana with flawless result. A moderate supply of shoots depends on the growth of suckers, thus giving rise to new individuals capable of producing an inflorescence [10]. It was a kind of culture that just wasn't worn in modern times, but has been practiced since primitive times as well.

Selection of Mother Plant

The mother plant should be healthy, free from disease and pests, especially viral diseases. The suckers should be selected from a healthy plant with a minimum height of 30 cm. Then, the extracted suction cup should be handled with care and implantation should be done as soon as possible. The selection of suckers follows by judging that of the plant: a wide bulb, narrow blades like leaves, an age of 2 to 4 months is preferable. Preferable suckers should weigh between 0.5 kg and 1.5 kg and the stem diameter should be around 2-6 inches. The leaves of mother plants are usually cut in the nursery (Figure 1).

Figure 1. Selection of a disease free plant.

Surface Sterilization

In the laboratory, the selected explants are soaked in detergents or disinfectants, viz. dettol for 60 minutes followed by washing under tap water for 15-20 minutes. The explants separated from the suckers are then treated in a surface sterilized laminar flow chamber. The explants are treated with mercuric chloride for 2 minutes and then washed 3 times with distilled water [10].

Preparation of Explant

The selected explants of the banana sucker are reduced into small pieces (0.5 to 1 mm in length). Removal of subsequent layers of suckers is necessary.

Culturing and Sub Culturing

The rejects are inoculated into the culture vessels and the vessels are maintained under optimal conditions (at temperature 280°C. and exposure to light for 16-18 hours) then observed for initiation, firing and rooting (Table 2). During this time, the greatest care is taken with the seedlings to avoid any kind of contamination and this cnsurcs a hcalthy futurc for all thc seedlings. After 4 weeks of inoculation, a first subculture is carried out. Its overall growth state is shown in Table 2. The blackened surface is scraped off and the growing apical shoot is cut to activate the helper buds. This cycle is repeated after 4 weeks and a second subculture is carried out.

Figure 2. Processing of sucker at TIU healthy mother plant of Grand Naine variety lab.

Primary Hardening

The small cups are 3/4 filled with visually fertile sterilized soil [10, 11]. The seedlings are taken out of the culture tanks and washed under tap water then placed in cups and covered with a polyethylene sheet to retain moisture and stored in optimal conditions. Liquid MS is administered for 4 days, followed by tap water. This process is continued for 14 to 15 days [11]. Healthy saplings are removed and the rest is discarded.

Secondary Hardening

Healthy young trees, obtained by primary hardening, are then transferred to polybags containing suitable substrates. Initially, these are kept in low light intensity shade nets at around 70% RH. By frequent application of water and NPK [3:3:1] fertilizer, the plants gradually harden. And finally, for further propagation, they are provided to farmers.

Figure 3. Inoculation of explants (21 days).

Figure 4. First subculturing at 21st day.

Result and Discussion

Table 2. Growth index and *in vitro* shooting and rooting characteristics of banana plantlets

Media	Date	Plant Inoculated	No. of Infected Plants	Explants Avg Size (cm) [Mean Value] (Figure 3)		Remarks
				Roots	Shoots	
MS Medium	05.02.18	50	10	Nil	Nil	No rooting or shooting were observed
	15.02.18	40	4	Nil	Nil	
	26.2.18	36	1	0.1	nil	Root initiation was started
	6.3.18	35	0	0.8	0.4	Root elongation and shoot initiation was seen
	16.3.18	35	0	1.4	1.4	Root and shoot elongation along with slight branching of roots
	26.3.18	35	0	3.18	4.51	Root and shoot elongated gradually
	15.4.18	35	0	8.74	21.9	Prominent root and shoot observed
Mean				2.84	7.05	
SD				3.49	10.05	
CV				122.6	142.53	

The result showed that different doses of BAP and NAA, i.e., 1.0, 2.0, 3.0, 4.0 and 5.0 mg/l and 0.1, 0.2, 0.25, 0.3, 0.35 respectively, influenced the growth of the shoot tips, as shown in (Table 3). After two weeks of culture for organogenesis, all cultures from all replications were refreshed on the same medium. On all the concentrations, the BAP at a concentration of 4 mg/bed, the NAA 0.25 mg/l gave the best response [10] which represents 95%. During cultivation period no. number of shoots gradually increases in all media supplemented by this concentration. After having undergone the whole process, the gradual developments of the samples are observed with the respective use of specific nutrients. The observation of the physical growth of the young tree is given in Table 2.

Figure 5. Primary hardening.

Figure 6. Secondary hardening.

Table 3. Effect of different concentration of plant growth regulators on micropropagation

Concentration [mg/l] of BAP + NAA in MS Medium	No. of Explants Inoculated	Percentage of Explants Regenerated	No. of Shoot Per Explant
1 + 0.1	15	25	1
2 + 0.15	15	45	2
3 + 0.2	15	60	2
4 + 0.25	15	95	3
5 + 0.3	15	85	2
6 + 0.35	15	45	1

Conclusion

Banana is a staple food and except that it has various other applications such as fodder, fiber, pigment, etc. Disease free high yielding somaclones of banana is very rare and is a challenge for the growers. So our effort is to produce these plantlets using simple protocol in MS with BAP and NAA for initiation, shooting and rooting of banana. The protocol raised in the present attempt could be used for commercial production of banana plantlets.

Acknowledgment

We wish to express our deep sense of gratitude to the chairperson of Techno India group, Sri Gautam RoyChowdhury Sir for his constant support and also for providing us the facilities which were needed for performing the research work in TIU laboratory. We also express our heartfelt thanks to Professor Dr. Nirmal Mandal, BCKV for his knowledge full support and guidance about the methodologies, without which completion of the research wouldn't have been possible. We also thank the respective authority of Techno India group viz Sri A. K. Roy and Sri Anit Adhikari for giving us constant support needed for the completion of our project. Finally we would like to thank all the team members and research scholars of BCKV for their constant support and enthusiastic help.

References

[1] Muhammad, A., Hussain, I., Saqlan Naqvi, S. M. and Rashid, H. 2004. "Banana plantlet production through tissue culture." *Pakistan Journal of Botany* 36(3):617-620.

[2] Ortiz, R. & Vuylsteke, D. R., 1996. "Recent advances in Musa genetics, breeding and biotechnology." *Plant Breeding Abstracts*. 66. 1355-1363.

[3] Kaemmer, D., Afza, R., Weising, K., Kahl, G., & Novak, F. J. 1992. "Oligonucleotide and amplification fingerprinting of wild species and cultivars of banana (Musa spp.)." *Bio/technology (Nature Publishing Company)*, 10(9), 1030-1035. https://doi.org/10.1038/nbt0992-1030.

[4] Mantell, S. H., J. A. Matthews, and R. A. McKee. 1985. *Principles of Plant Biotechnology: An Introduction to Genetic Engineering in Plants*. Oxford: Blackwell Scientific Publications.

[5] Arinaitwe, G., Rubaihayo, P. R. and Magambo, M. J. S., 1999. "Effects of auxin/cytokinin combinations on shoot proliferation in banana cultivars." *African Crop Science Journal*, 7(4), pp. 605-611.

[6] Bairwa, K. C., Ajay Singh, A. Jhajaria, H. M. Singh, B. K. Goyam, M. Lata and N. K. Singh. "Conventional sucker with tissue culture banana production in Central India- A case study." *African Journal of Agricultural Research* 10 (2015): 1751-1755.

[7] Buah, J. N., Eric Oppong Danso, K. J. Taah, E. A. Abole, E. A. Bediako, J. B. K. Asiedu and Richard Ogirma Baidoo. "The effects of different concentrations cytokinins on the *in vitro* multiplication of plantain (Musa sp.)." *Biotechnology (faisalabad)* 9 (2010): 343-347.

[8] Ahmed, S., Akash Sharma, A K Singh, Vinod Kumar Wali and Preeti Kumari. "*In vitro* multiplication of banana (Musa sp.) cv. Grand Naine." *African Journal of Biotechnology* 13 (2014): 2696-2703.

[9] Rahman, S., Nirupam Biswas, M. Hassan, G. Ahmed, A. Mamun, R. Islam, Moniruzzaman and Enamul Haque. "Micro-propagation of banana (Musa sp.) cv. Agnishwar by In vitro shoot tip culture." *International Research Journal of Biotechnology* 4 (2013): 83-88.

[10] Kalimuthu, K., Murugesan Saravanakumar and Rajendran Senthilkumar. "*In vitro* micropropagation of Musa sapientum L. (Cavendish Dwarf)." *African Journal of Biotechnology* 6 (2007): n. pag.

[11] Avvaru, S. "Plant tissue culture of banana in laboratory." *Research and Reviews: Journal of Botanical Sciences* 5 (2016). 54-62.

About the Editors

Dr. Sanket Joshi

Dr. Sanket Joshi (https://orcid.org/0000-0001-7746-9905) is a Deputy Director, Oil & Gas Research Centre, and an Application Specialist, at Central Analytical and Applied Research Unit, Sultan Qaboos University, Oman. He holds BSc and MSc degrees from Sardar Patel University, India, and a PhD degree from M. S. University of Baroda, India – all in Microbiology.

Dr. Joshi has 16 years of academic teaching and research experience, and 4 years of industrial R&D experience, in India and Oman. While working in Indian pharma companies, he undertook several turnkey projects (both upstream and downstream) on antibiotics (β-lactams, macrolides, glycopeptides), antidiabetic drug (for type II diabetes), and Co-enzyme Q10. His current research interests encompass: Energy (*In-situ/ex-situ* microbial enhanced light/heavy oil recovery; chemical enhanced oil Recovery; biofuels); Microbial products (biosurfactants, biopolymers, R&D and scale-up); and Environmental bioremediation (crude oil pollution; analysis, mitigation and control of souring by Sulfate reducing bacteria; HPAM contaminated sites). Those research projects are multidisciplinary, involving both biotechnology and engineering disciplines. Since he joined SQU, his research team has received over 4.28 million USD in research grants and service contracts, and established bioproducts for petroleum and environmental applications. He has 154 scientific publications as papers in international journals, book chapters, conference proceedings, international books, and one Australian Innovation Patent application to his credit.

Dr. Joshi serves as an Academic/Associate Editor for some of the highly reputed journals: *Frontiers in Microbiology*, *PeerJ*, *Ecotoxicology* (Springer), *Petroleum Science and Technology* (Taylor & Francis, UK), *3 Biotech* (Springer), *Applied Biochemistry and Biotechnology* (Springer), and *Open Biotechnology Journal*. He is book series editor of Advances in Biotechnology and Bioengineering (Elsevier Inc.); Guest editor for *Frontiers in Microbiology*, *Sustainability*, *Scientifica*, *Biocatalysis and Agricultural Biotechnology*, and *Open Biotechnology Journal*; and a recognized reviewer for more than 51 international journals (*Elsevier*, *Springer Nature*, *T&F*, *Wiley/AOCS*, *RSC*, *ACS*, *Hindawi*, *DE Gruyter*, *MDPI* and others).

Dr. Joshi served as an intermediate guide for 30 undergrad/grad students and 3 PhD students, in India and in Oman. He is also an active member of MEOR research team at SQU since 2009, and established a state of art research and service laboratory at SQU.

Dr. Joshi was awarded as "NRI Senior Scientist Award", from Microbiologists Society, India, during academic year, 2019-2020. He was ranked 27th in SQU among top 500 authors (based on number of citations 2010-2019, as per SciVal-SCOPUS/Elsevier), with 3010 citations, *h-index* of 26, and *i10* index of 44 Google Scholar).

Professor Susmita Mukherjee

Professor Susmita Mukherjee (https://orcid.org/0000-0002-0860-5018) is Professor and Head of the Department of Biotechnology, University of Engineering & Management, Kolkata, India. She did her Graduation and Post-graduation in Zoology then started working on the ecological aspects of wastewater fisheries.

Passionate to learn and work on Environmental Biotechnology, Professor Mukherjee has been working in this field for more than 10 years towards preservation of ecosystems. She has done extensive work on the effect of ecological factors on sewage fed fisheries of East Kolkata Wetlands, West Bengal, India. Her doctoral work was on the wastewater management of Kolkata city. Her research interest is in environmental toxicology and waste bioremediation primarily of land and aquatic systems. She also worked on the algal bioremediation of wastewater, understanding the effectiveness of the algal system involving different variations compared to other prevailing indigenous systems towards better management of wastewater problem of Kolkata. She has worked on soil arsenic remediation in the Atghara area of West Bengal.

Professor Mukherjee has authored many journals and conference papers. She also has worked as an editor of books published by Springer and Taylor & Francis, UK. Her present project is on the decontamination of a landfill site.

Dr. Moupriya Nag

Dr. Moupriya Nag (https://orcid.org/0000-0002-1365-5722) is presently working as Assistant Professor in the Department of Biotechnology, University of Engineering & Management, Kolkata, India. Her Postdoctoral work was focused on exploring the aggregation kinetics and mechanism of action of peptides associated with neurodegenerative diseases in presence of various membrane surfaces with solution NMR. It was a rewarding experience as she learnt various state of the art techniques used in a molecular biology lab.

Dr. Nag's research interests include single molecule Biophysics of protein unfolding/refolding and protein aggregation kinetics and its mechanism of action especially the amyloid proteins in neurodegenerative disorders. Recently, her research focuses on functional amyloids in bacteria and their relation in forming biofilms. Her work deals with developing novel antimicrobial and anti-biofilm compounds from natural sources including plants, microbes and green synthesized nanoparticles.

Dr. Nag has more than five years of teaching experience in Biophysics, Bioinformatics, Bioprocess and Biotechnology, Enzyme Technology, Computing, Molecular Modeling and Drug Design. She has also published many research articles in peer-reviewed international journals and authored or co-authored numerous book chapters and books. She is a member of many scientific societies like Indian Biophysical Society and Indian Science Congress Association, and an Associate Editor for *Applied Biochemistry and Biotechnology*, Springer.

Index

C

O

P

R

S